Fourth Edition

COMPUTATIONAL FLUID DYNAMICS VOLUME III

KLAUS A. HOFFMANN

STEVE T. CHIANG

A Publication of Engineering Education System™, Wichita, Kansas, 67208-1078, USA

www.EESbooks.com

ISBN 0-9623731-6-8
First Print: August 2000

This book is typeset by Jeanie Duvall dba SciTech Computer Typesetting of Austin, Texas.

To obtain information on purchasing this or other texts published by EES, please write to:

Engineering Education System™
P.O. Box 20078
Wichita, KS 67208-1078
USA

Or visit:

www.EESbooks.com

CONTENTS

Appendices

PREFACE

The fundamental concepts and development of computational schemes for the solution of parabolic, elliptic, and hyperbolic equations are established in Volume I and are extended to the Navier-Stokes equations in Volume II. The primary goal in the third volume is to review the fundamentals of turbulence and turbulent flows and to extend the governing equations and numerical schemes developed in Volume II to include turbulence.

This volume begins with the basic definitions and concepts in turbulence and turbulent flows. Subsequently, the modification of the governing equations and numerical schemes is introduced. There are three approaches by which turbulent flowfields may be computed. The first approach is based on the averaged Navier-Stokes equations either in the form of Reynolds-Averaged Navier-Stokes (RANS) equations or Favre-Averaged Navier-Stokes (FANS) equations. These formulations, along with several turbulence models and numerical considerations for the solution of equations, are presented in Chapter 21. The second and third approaches are the Large Eddy Simulations (LES) and the Direct Numerical Simulations (DNS), which are presented in Chapter 23. Since typical computations involved in turbulence and, in particular, in DNS, require higher-order schemes such as compact finite difference formulations, these formulations are introduced in Chapter 22.

Finally, a computer code based on the RANS equations and several turbulence models have been developed and included in the text *Student Guide for CFD–Volume III.*

Again, our sincere thanks and appreciation are extended to all individuals acknowledged in the preface of the first volume. Thank you all very much for your friendship and encouragement.

Klaus A. Hoffmann
Steve T. Chiang

Chapter 20
Turbulence and Turbulent Flows

20.1 Introductory Remarks

The vast majority of applications in fluids involves turbulence. Everyday experiences such as fluid flow in a pipe, flow over an airfoil, flow processes in combustion, paint spraying, etc. will exhibit a disorderly complex motion defined as turbulent flow. Since the majority of fluids involve turbulence, turbulence flow control becomes an important aspect of any design process involving fluids. In some applications it is advantageous to promote turbulence in order to achieve certain design goals. Introduction of dimples on a golf ball is an example whereby enhancing transition to turbulent flow reduces flow separation region, and therefore total drag is decreased.

Due to the complexity of turbulent flow and difficulties in its understanding and physical interpretation, introduction of a unique conceptual model is difficult at best. However, there are certain characteristics and properties associated with turbulence which have been shown experimentally and most recently numerically, which are used commonly in describing turbulence and turbulent flows. A brief description of the physics of turbulence, its generation and development, and a conceptual model is presented in this chapter. However, keep in mind that this description is not unique and others are provided in literature. The conceptual model described in this chapter is provided to shed some light into the world of turbulence. A review article by Robinson [20.1] is an excellent source for description of several conceptual models for turbulence and is highly recommended. Following the description of Ref. [20.1], a typical conceptual model is described to identify some of the physical characteristics of turbulent flows. Furthermore, several categories of turbulent flows are described and some well-established properties of turbulent flows are summarized. Finally, the three categories of computational approaches associated with turbulent flows are described.

A turbulent flow may be defined as a flow which contains self-sustaining fluctuations of flow properties imposed on the main flow. There are several factors which may cause an originally laminar flow to transition to turbulence. The fundamental quantity in describing transition to turbulent flow is the Reynolds number. For example, for internal pipe flow a transition to turbulent flow can be achieved at around a Reynolds number of 2300 and for a boundary layer flow over a flat plate at a Reynolds number of around 300,000 $\sim$ 500,000. Obviously, several factors affect transition to turbulent flow including freestream turbulence, pressure gradient, heat transfer (cooling or heating), surface roughness, and surface curvature. Turbulent flows can be broadly categorized into three groups: boundary layers, shear layers, and grid-generated turbulent flows. The first two categories are most common and will be the focus of our study.

The first category of turbulent flows is the wall-bounded flows. In these types of turbulent flows, the majority of turbulent kinetic energy is produced near the wall region. Furthermore, the small scale eddies which are dominant in the near wall region are more organized and have similar structure. The wall-bounded turbulent flows can be further subcategorized as turbulent boundary layer or fully developed turbulent flow. A turbulent boundary layer is simply bounded by a wall and a free stream, whereas a fully developed flow is bounded by surfaces, for example, flow in a channel or flow in a pipe.

The second category of turbulent flows is the shear layer. This type of turbulent flow grows in the streamwise direction and typically develops universal characteristics, which may be considered as self-preserving. The turbulent shear layers may be subcategorized into three groups as free shear layers, jets, or wakes. A typical mean flow velocity profiles for the three categories of shear flows are shown in Figure 20-1.

The flow field for the flows described above would of course be considerably more complex if the flow is supersonic. Such a flowfield will include expansion waves, compression waves, shock waves, their interactions with each other, and with the shear layer. A typical flowfield for the supersonic wake flow is shown in Figure 20-2.

The third category of turbulent flow is the grid generated turbulent flow, which can be produced by passing an initially uniform and irrotational flow through a grid composed of rods. The vortices generated by the rods interact together and degenerate into turbulence. The grid-generated turbulence is typically isotropic, that is, it has a preferred direction. In general, most turbulent flows are anisotropic. However, the assumption of isotropy is used in many applications.

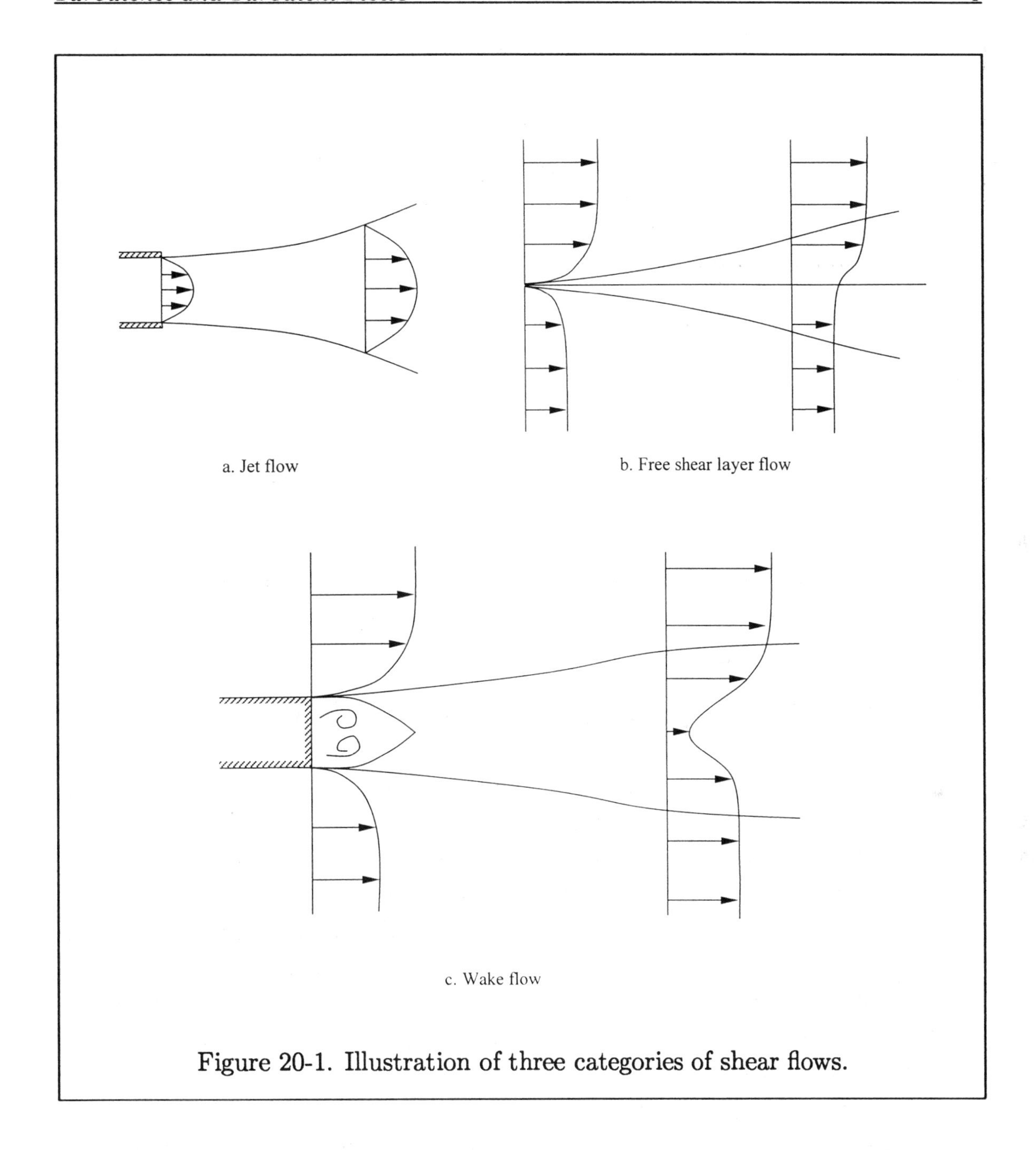

Figure 20-1. Illustration of three categories of shear flows.

The simple turbulent boundary layer over a flat plate and a free shear layer will be used to introduce conceptual models. Several processes associated with the production of turbulence will be described in this chapter. Before attempting to introduce a conceptional model, a brief description of transition is warranted, which is given next. Subsequently, some fundamental concepts and definitions are reviewed leading to introduction of a conceptual model.

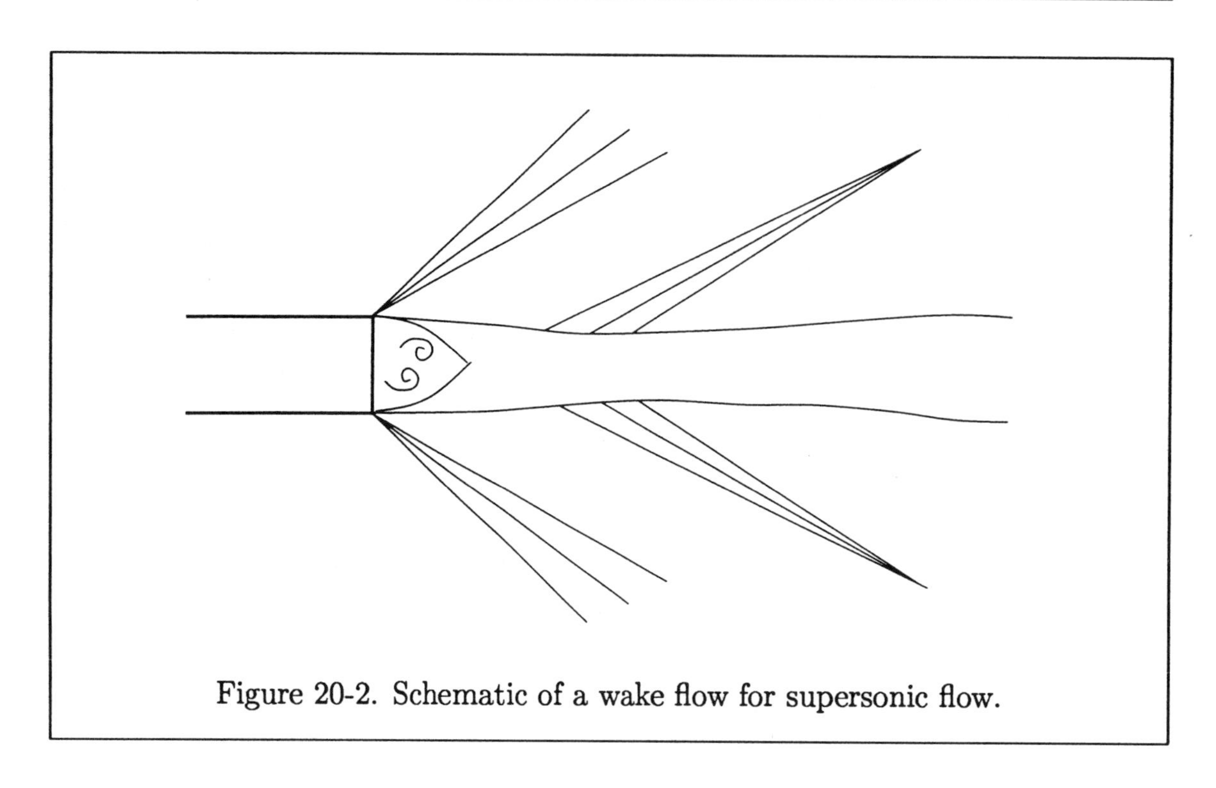

Figure 20-2. Schematic of a wake flow for supersonic flow.

20.2 Fundamental Concepts and Definitions

Before attempting to describe a conceptual model and mathematical models, it is necessary to review several fundamental concepts, physical observations, definitions, and terminology related to turbulence and turbulent flows. This goal will be accomplished primarily by considering turbulent boundary layers.

A turbulent boundary layer is commonly divided into several regions, where within each region specific turbulence behavior can be identified. A very thin region near the surface is referred to as viscous sublayer. The outer region where the flow is turbulent is called the fully turbulent zone. A region, which connects these two zones, is called the buffer zone.

The viscous sublayer has been referred to as the laminar sublayer as well. However, recent experimental and numerical investigations have shown that in fact this region is not laminar. Similarly, the description of buffer zone, which in the past has been referred to as a region of transition from laminar to turbulent, is not accurate. The three regions across a turbulent boundary layer defined above is used extensively in description of turbulent flows as well as in development of turbulence models. Further description of each region based on certain characteristic of turbulence development will be presented throughout this section.

Another categorization of a turbulent boundary layer is to divide it into inner and outer regions. The inner region includes the viscous sublayer, buffer zone, and part of the fully turbulent zone. The remaining part of the turbulent boundary layer is considered to be in the outer region.

The division of a turbulent boundary layer to several regions defined above is commonly identified by definition of a non-dimensional velocity u^+ and normal to the surface spatial coordinate y^+. These quantities are defined as

$$u^+ = \frac{u}{u_\tau}$$

and

$$y^+ = y\frac{u_\tau}{\nu}$$

where u_τ is know as the friction velocity given by

$$u_\tau = \sqrt{\frac{\tau_w}{\rho}}$$

and τ_w is the wall shear stress. Now, the various regions are identified as

$$y^+ < 2 \sim 8 \Rightarrow \text{viscous sublayer} \tag{20-1}$$

$$2 \sim 8 < y^+ < 2 \sim 50 \Rightarrow \text{buffer zone} \tag{20-2}$$

$$y^+ > \sim 50 \Rightarrow \text{fully turbulent zone} \tag{20-3}$$

and

$$y^+ < 100 \sim 400 \Rightarrow \text{inner region} \tag{20-4}$$

$$y^+ > 100 \sim 400 \Rightarrow \text{outer region} \tag{20-5}$$

Observe that the division between each region in terms of y^+ is approximate, and different values have been used by investigators to identify each region. However, the range given above is the most common. The various regions of turbulent boundary layer are summarized in Figure 20-3.

20.3 Introduction to Transition

Majority of flows begin as orderly fluid motion known as laminar flow. As the flow Reynolds number is increased, instabilities within the boundary layer are generated. Subsequently, several physical phenomena are developed which eventually will lead to transition from laminar flow to a disorderly and random fluid motion known as turbulence. Several factors may enhance or delay the transition process. Examples of such enhancement mechanism are surface roughness, heat transfer, pressure gradient, and freestream turbulence.

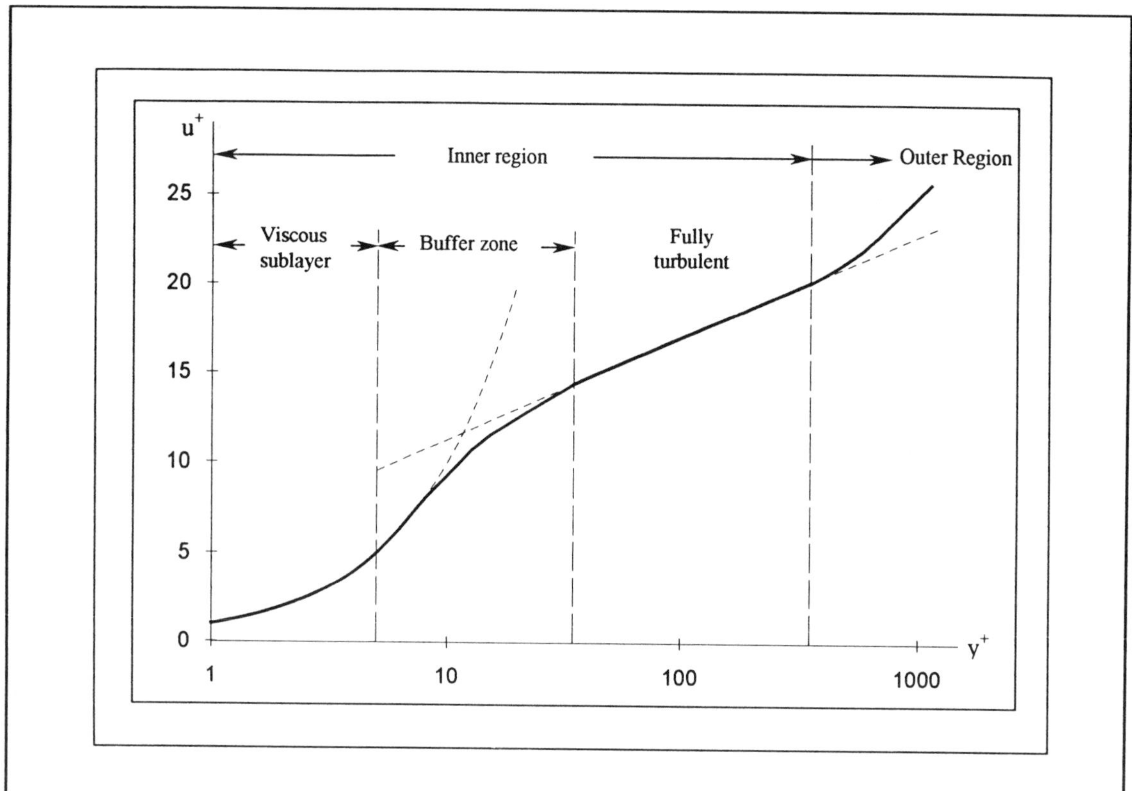

Figure 20-3. Typical nondimensional velocity profile for a turbulent flow over a flat plate.

In general then, transition is defined as a change where instabilities within a laminar boundary layer are initiated. Subsequently several processes such as vortex stretching, vortex breakdown, and formation of turbulent spots take place, and the flow becomes fully turbulent. The entire changes and processes which occur over time and space is called transition. Transition process is a complex set of events which is extremely difficult to analyze or develop theoretical models for its prediction. With the exception of direct numerical simulation (DNS) which is currently very limited in applications, no theoretical model for prediction of turbulence and its development exists. The only available approach is semi-empirical models developed based on experimental data. Our objective at this point is to briefly review some of the physical phenomena, which occur during transition. This goal is accomplished first by visiting the small disturbance stability theory, because the theory can predict onset of instabilities for idealized cases. Subsequently a brief description of the processes, which completes the process of transition, is provided. We will proceed with description of several relevant stages without detailed mathematical work. A description of stability theory is given first, followed by description of processes, which completes transition.

20.3.1 Stability Theory

In this section the stability theory is reviewed. The mathematical details are deleted; instead the procedures to obtain the governing equations and the conclusions drawn from the solution of the equations are emphasized. The mathematical details may be found in texts by White [20.2], Panton [20.3], and Landahl and Mello-Christensen [20.4].

The governing equation of stability is primarily derived for an incompressible flow with constant properties. Assume a vector Q to represent the flow variables and Q' to represent perturbations (disturbances) imposed on the flow. The flow variable $Q + Q'$ is substituted into the Navier-Stokes equations, and the original equations written in terms of Q are subtracted from it. The equations thus obtained involve the perturbation variable Q' as the unknown. The resulting equations will be non-linear in perturbation properties Q', thus, typically the equations are linearized by neglecting the higher order terms in Q'. This is in fact a reasonable assumption, because the disturbances are assumed to be small. Now, if for a given problem the disturbance Q' will decay, the problem is stable. On the other hand, if the disturbance Q' grows with time, the problem is unstable. The governing equation of stability may be solved numerically for a wide range of applications. Traditionally, analytical solutions of stability equation have been achieved for special cases where additional assumptions had to be imposed to reduce the equation. One such solution, which is relevant to boundary layer flows, is reviewed at this point.

The problem of interest is the nearly parallel viscous flow. With this assumption, the stability equation is reduced to a single ordinary differential equation. Typically a disturbance in the form of an exponential function is introduced and the resulting equation is known as the Orr-Sommerfeld equation.

At this point, the fundamentals and procedures for obtaining the governing equations of stability have been reviewed. Again deleting the mathematical details or the solution of Orr-Sommerfeld equation for parallel flow, several observations regarding the solution of the equations are presented.

Typical solution of the Orr-Sommerfeld equation is given graphically in Figure 20-4 for a Blasius boundary layer flow. The horizontal axis represents Re_{δ^*} where $Re_{\delta^*} = (\rho u \delta^*)/\mu$, and the vertical axis is $\alpha\delta^*$, where α is the wave number in the x-direction and δ^* is the displacement thickness. A curve defining the boundaries of the unstable waves which separates the stable and unstable regions is called marginal (or neutral) stability curve. A point on the marginal stability curve corresponding to lowest Re_{δ^*} is called *the critical point.* Within the curve of marginal stability is the range of unstable wave numbers. These unstable waves are known as *Tollmein-Schlichting waves*, which represent the first indication of instability.

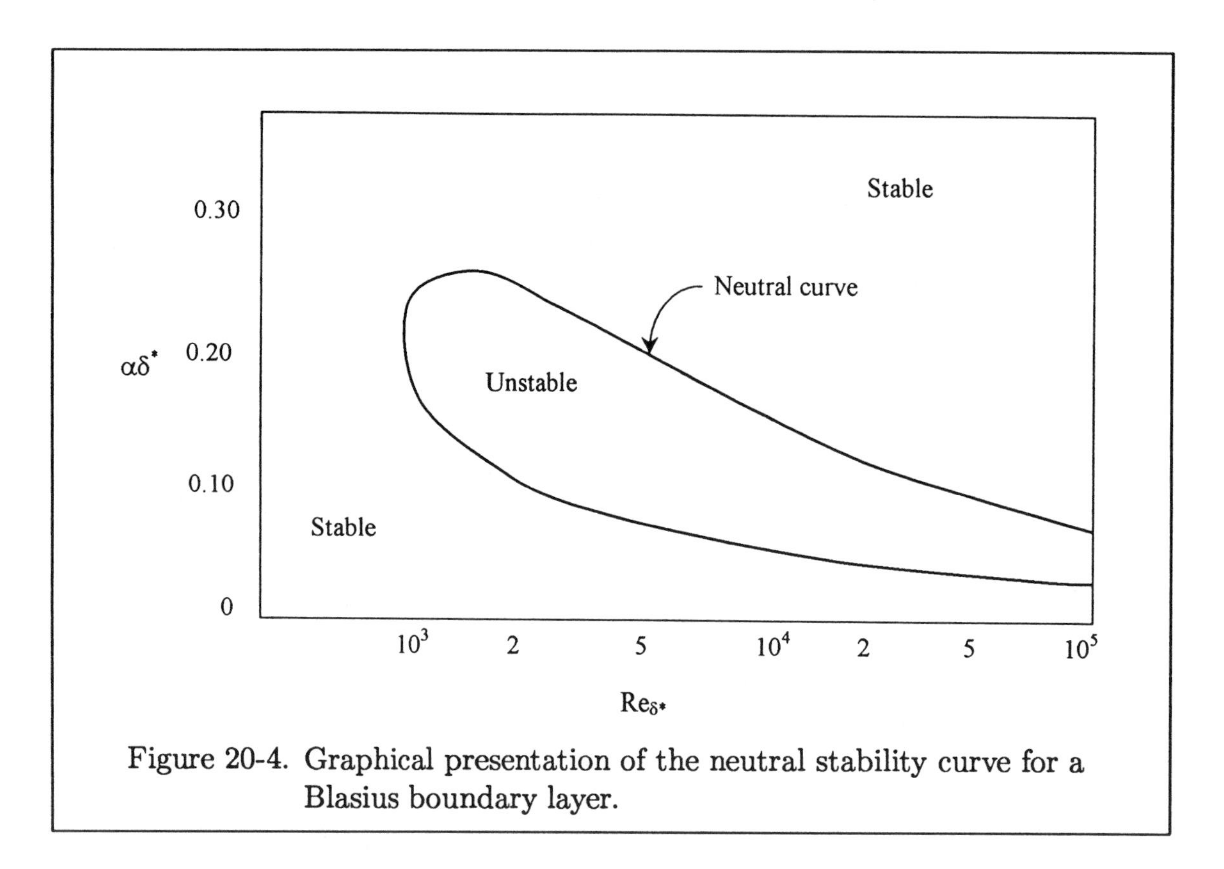

Figure 20-4. Graphical presentation of the neutral stability curve for a Blasius boundary layer.

An interesting result obtained from the stability equation applied to Blasius boundary layer reported by Squire [20.5] is that when considering a three-dimensional disturbance, the corresponding two-dimensional waves are more unstable. This conclusion is important in that two-dimensional analysis can be used to determine the limit of stability, which is also valid for the three-dimensional case.

The following important statement concludes the discussion on stability. The stability theory can predict lowest Reynolds number at which instability can be initiated. However, note that this is not the location of transition to turbulence. Occurrence of instability is only the first indication of a process to transition. In fact, typically transition occurs at a distance of 10 to 20 times the distance of x_{critical}. Furthermore, it should be noted again that several factors affect transition. For example, surface roughness, pressure gradient, freestream turbulence, surface curvature, and heat transfer have a dominant effect on the process of transition. The stability analysis detailed here and a description of transition to follow is for a smooth plate with no external factors imposed. The purpose at this point is to present a descriptive model for transition in its simplest form. There are still a very large number of physical phenomena with regard to transition and turbulence, which are not completely understood.

20.3.2 Transition

Based on the discussion of stability just completed, it is well known that at some critical Reynolds number instabilities in the form of Tollmein-Schlichting waves which are two-dimensional are developed. Shortly thereafter, the unstable waves become three dimensional and nonlinear effects come into play. As mentioned previously, these instabilities by themselves do not define transition. In fact several mechanisms and processes subsequent to initial instabilities must take place to complete the process of transition from laminar to turbulent flow. Once the three-dimensional instabilities are set in, the nonlinearity effects and interaction with the mean flow, result in changes in the mean flow velocity and spanwise stretching of vorticity. In addition internal shear layers are formed which are highly unstable. Subsequently a breakdown process sets in where the stretched vortices breakdown to smaller vortices with random frequencies and amplitudes, producing localized regions of high shear and intense fluctuations in the shear layer. As a result, random and disorderly motion termed turbulence is generated which travels downstream and grows in coverage. This process occurs in localized regions surrounded by laminar flow and is called turbulent spot. Eventually the number of spots is increased in the streamwise direction and the flow becomes fully turbulent. This in turn completes a process, which is termed transition. This brief description of transition is based on a flow over a smooth flat plate. The transition process can be enhanced or delayed by several factors identified earlier. A graphical presentation of transition process as proposed by White [20.2] is shown in Figure 20-5.

20.4 Conceptual Model for Turbulent Flows

A model developed to describe the physics of a problem is called conceptual model. A conceptual model for turbulent flow may be developed based on experimental observations and complemented by numerical observations. At this point in time, there are several aspects of turbulent flows, which are poorly understood. As a result there is clearly a need for improvement of proposed conceptual models. Furthermore, due to uncertainties and the lack of clear understanding and interpretation of available data, there are different and sometimes contradictory explanations of events related to turbulence. With these remarks in mind, it is then obvious that several conceptual models for turbulent flows can be and have been proposed. These models are updated as we gain more understanding about turbulent flows and it is anticipated that they will eventually converge to a widely acceptable model. For the time being though, there are several conceptual models proposed by investigators. The objective here is not to review all the conceptual models. Instead the goal is to review and familiarize the reader with some of the termologies used in description of turbulence.

The flow very near the surface in the sublayer and buffer regions moves as streaks of low and high-speed fluid. Why these streaks form is still not well understood, although Blackwelder [20.6] has suggested that 0.25 μm (10^{-5} in) size dimples (local depressions) are sufficient to produce the critical conditions necessary for vortex formation. Associated with the low and high-speed streaks are counter rotating streamwise vortices, as shown in Figure 20-6.

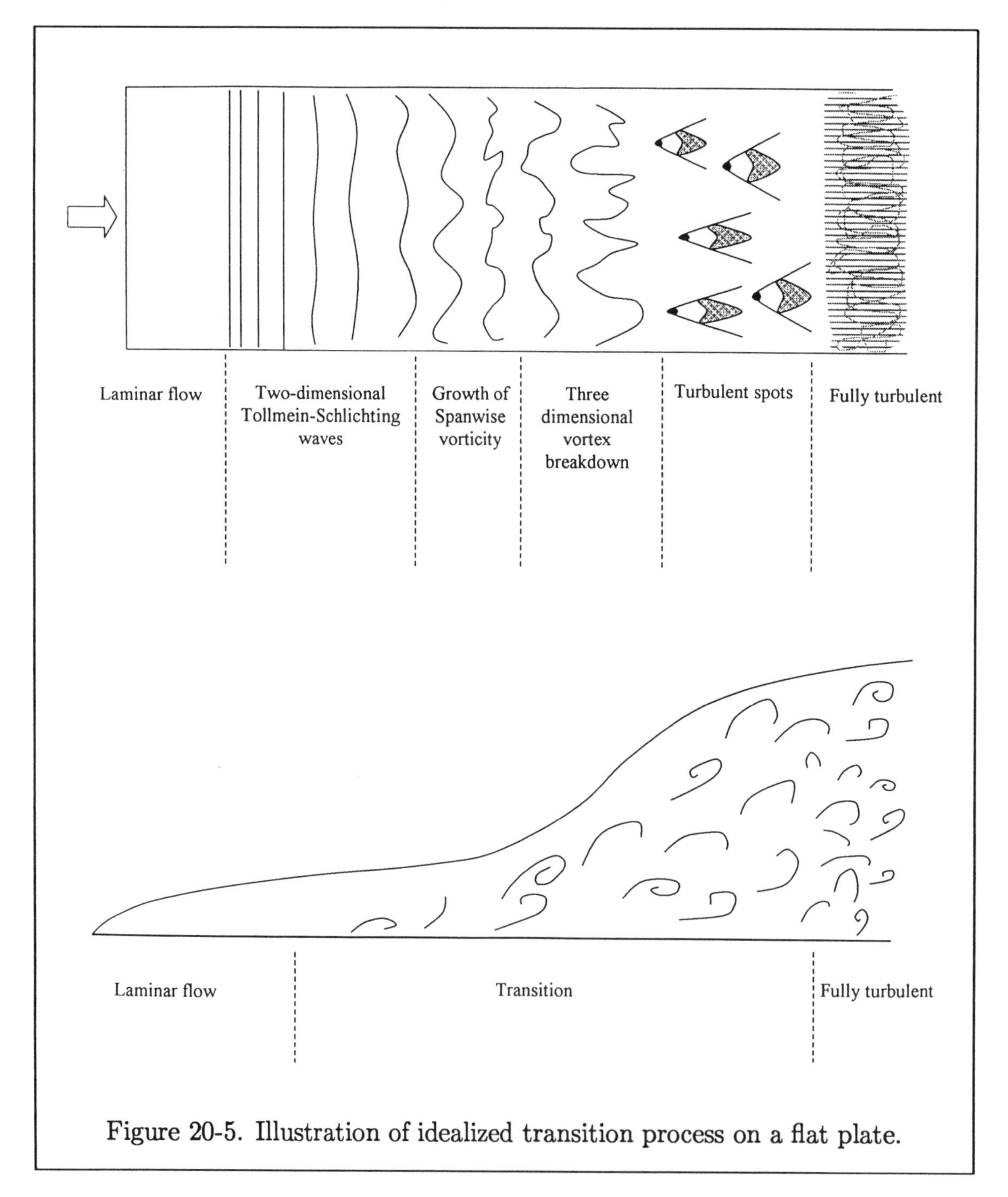

Figure 20-5. Illustration of idealized transition process on a flat plate.

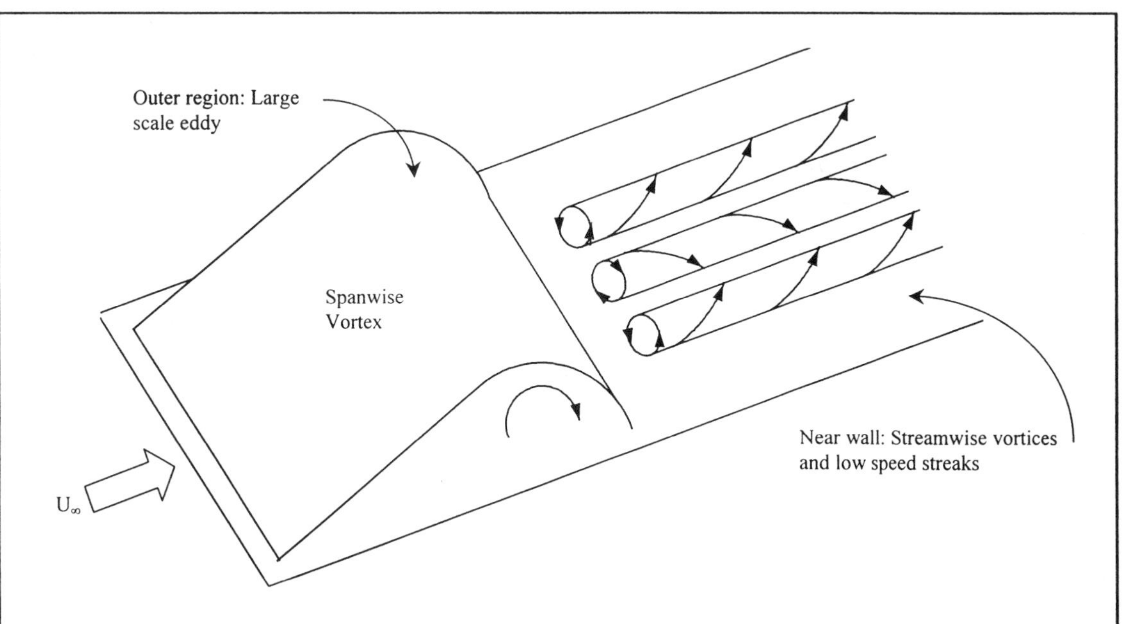

(a) Structure of turbulent boundary layer [20.7]

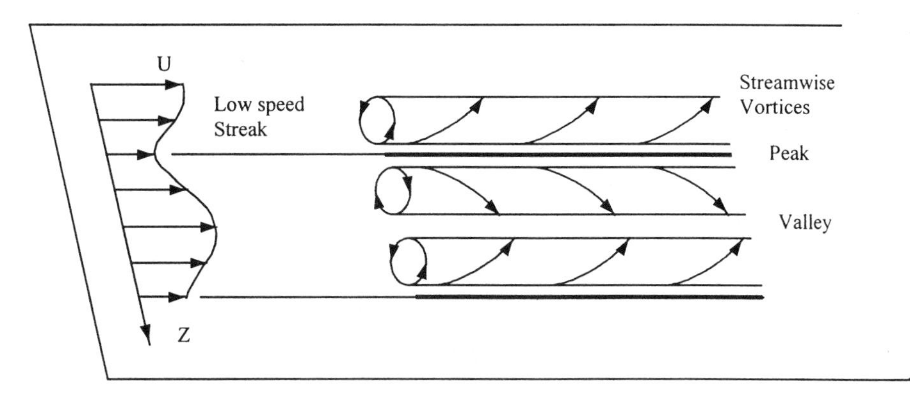

(b) Near wall region has low and high speed streaks and the counter-rotating stream-wise vortices are associated with these streaks. [20.7]

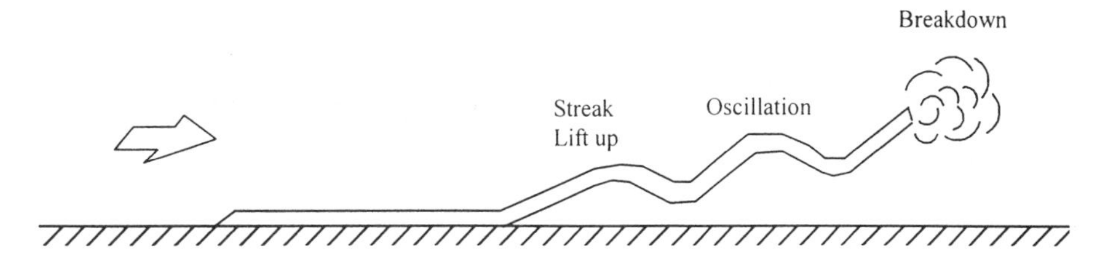

(c) low speed streak lift up off the wall leading to breakdown, known as bursting. [20.7]

Figure 20-6. Conceptual model for Turbulent boundary layer.

Intermittently, the low speed fluid is ejected outward accompanied by inrush of high-speed fluid. The ejection of the low speed flow, which is accompanied with oscillation and breakup, is termed bursting. It is observed that majority of turbulence energy is produced in the sublayer/buffer zone regions [20.8].

It is well known that the edge of a turbulent boundary layer is irregular with peaks and valleys on the scale of boundary layer thickness. Freestream fluid is entrained into the turbulent boundary layer at the irregular interface. Furthermore, it is observed that the flow in the outer region of the boundary layer under the bulges develops rotational spanwise eddies [20.9]. This outer region large-scale eddy is illustrated in Figure 20-6.

There is ample evidence that the turbulence in the inner region is self-sustaining and that the outer layer has some effect on the near wall production process [20.10]. Obviously, there is an interaction between the inner and outer regions in that there is a transfer of mass, momentum, and energy from the outer layer to the inner layer and vice versa.

Vortical structures have also been used extensively in description of conceptual models of turbulent boundary layers. A brief and limited discussion of voritcal structures is described in this section. The purpose again is to familiarize the reader with some of the proposed models. For a comprehensive review Reference [20.1] is highly recommended. The proposed conceptual model populates the turbulent boundary layer with various vortices, which are the elements in the turbulence production and dissipation as well as transport of mass and momentum between the inner and outer layers. One of the most common vortical structures used is the hairpin vortex (in high Reynolds number region) and the horseshoe vortex (in low Reynolds number region) within the inner region. These vortices are inclined in the downstream direction at relatively small angles from the wall, in the range of $10^\circ \sim 50^\circ$. The formation of these vortices can be visualized as lifting of the streamwise vortices, which are stretched and deformed to a horseshoe vortex. As discussed previously, spanwise vortices or large eddies are developed in the outer region.

20.5 Production, Diffusion, and Dissipation of Turbulence

Any proposed mathematical model must be based on correct representation of the physical processes involved in the problem under consideration. For example, the Navier-Stokes equations include unsteady, convective, and diffusive terms. Similarly in developing a turbulence model, physical processes involved must be identified and must be included in the proposed mathematical model. These processes include unsteadiness, convection, production, diffusion, and dissipation of turbulence. A basic description of production, diffusion, and dissipation of turbulence is to follow.

Turbulence is produced by different processes depending on the physics of the problem. For example, turbulence is produced in the inner region of a turbulent boundary layer by a process called bursting or one may argue the production is the result of formation of hairpin vortices. The turbulent eddies formed in the inner region are typically of small scale. On the other hand, turbulence production in the outer region and formation of eddies are of larger scales. Processes also occurs where smaller eddies may grow and become larger eddies and vice versa where the larger eddies become smaller and smaller. Thus, one aspect of any turbulent flow is production of turbulence, which is commonly referred to as production.

The second aspect is diffusion. Recall that mass, energy, and momentum are transported within a fluid domain due to random motion of molecules, given rise to mass diffusion, thermal diffusion, and viscous diffusion. It can be argued that due to the random motion of eddies in a turbulent flow, there is a transport of fluid properties from one region of the flow to another region. This process tends to increase the mixing process in the fluid and is referred to as eddy diffusion or simply as diffusion. Finally, recall that eddies of different scales are formed in a turbulent flow. The largest scale in a turbulent boundary layer has an upper limit of boundary layer thickness. Similarly, there is a lower limit beyond which the small-scale eddy is self-destructive. Of course, molecular viscosity is the primary factor. Note that as smaller and smaller eddies are formed, due to their large velocity gradient, the viscous forces become considerable and eventually the smallest eddies are destroyed. Furthermore, due to the large velocity gradient associated with smaller eddies; they dominate the energy dissipation process in the flow. This process of energy dissipation and destruction of small eddies is known as dissipation.

20.6 Length and Time Scales of Turbulence

It is obvious at this point that there is a wide range of scales associated with turbulent flows. The small and large eddies developed in a turbulent flow each possess certain characteristics which are important if one attempts to develop prediction methods for turbulent flows. For example, in direct numerical simulations it is imperative that the size of spatial grid be sufficiently fine in order to accurately capture the essence of small scale eddies.

Recall that most of the energy in a turbulent flow is contained in large eddies due to their intense velocity fluctuations. Furthermore, the lifespan of the large-scale fluctuations is relatively long. On the other hand, the energy associated with smaller eddies is smaller due to their smaller velocity fluctuations and their life span is considerably shorter. Furthermore, recall that kinetic energy is transferred from larger eddies to smaller and smaller eddies the so called cascading process. Eventually, the kinetic energy is dissipated into heat due to the action of molecular viscosity at the scale of smallest eddies.

Now consider a turbulent boundary layer with a freestream velocity of U_0 and a boundary layer thickness δ. Define a turnover time for the large eddies by δ/U_0. The energy of the large eddy is proportional to U_0^2. Hence, the rate of dissipation ϵ would be proportional to U_0^3/δ.

The scales of turbulence can be easily established for smallest eddies if one uses Kolmogorov universal equilibrium theorem which states that the rate of transfer of energy from larger eddies to smaller eddies is approximately equal to the dissipation of energy to heat by the smallest eddies. Therefore, the primary parameters for smallest eddies are the mean dissipation and kinematics viscosity. Hence, one can use dimensional argument to establish the following length (η), velocity (v), and time (τ) scales,

$$\begin{aligned}
\eta &= \left(\frac{\nu^3}{\epsilon}\right)^{1/4} \\
v &= (\nu\epsilon)^{1/4} \\
\tau &= \left(\frac{\nu}{\epsilon}\right)^{1/2}
\end{aligned}$$

which are known as the Kolmogorov scales of length, velocity, and time. The Kolmogorov length scale η is then a measure of smallest eddies which can withstand the damping effect of molecular viscosity. It is important to note that the smallest length scale of turbulence that we have defined as the Kolmogorov length scale is still much larger than the mean free path of molecules, the primary requirement of continuum model. Therefore, the continuum model is commonly used for turbulent flows.

It is appropriate at this time to review the concept of isotropy, which is used extensively in turbulent flows, and it has significant implications for turbulence models and large eddy simulation. Isotropy generically implies invariance with respect to orientation. The assumption of local isotropy is used for small turbulence scales. Experimental and theoretical considerations indicate that this assumption is reasonable under certain circumstances. Nonetheless, the assumption of local isotropy is used extensively. The implication of this assumption on the accuracy of solutions and alternate procedures are currently under intense investigations.

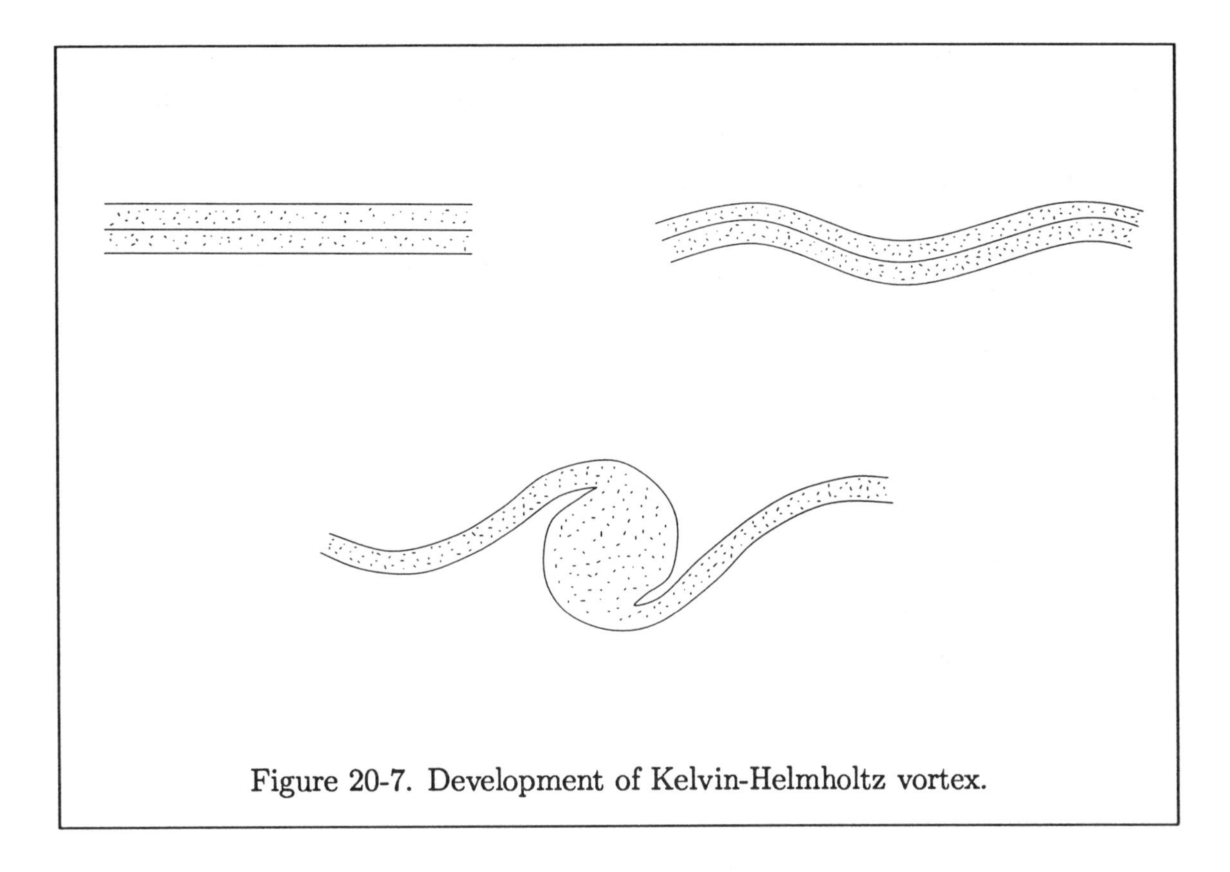

Figure 20-7. Development of Kelvin-Helmholtz vortex.

20.7 Free Shear Layer Flows

The general characteristics of turbulent free shear layers are reviewed in this section. Consider two flows at different velocities at the upper and lower surfaces of a splitter plate. Once the flows pass the trailing edge, instabilities known as Kelvin-Helmholtz instability is developed at the interface. Skematic of this process and formation of Kelvin-Helmholtz vortex is shown in Figure 20-7. Based on the stability analysis discussed previously, it can be shown that the primary factor in the development of these instabilities is governed by inviscid character of the flow. That is, molecular viscosity has little influence in the development of these instabilities as long as molecular viscosity is small. Recall that on the contrary, the primary factor in the development of wall bounded instability was molecular viscosity. Once the Kelvin-Helmholtz instability is initiated, due to the roll up and pairing, Kelvin-Helmholtz vortices are formed which are carried downstream. These large scale eddies are very regular and possess certain degree of spatial organization. Due to their regular formation, they are referred to as coherent structures. These large coherent structured eddies which encompass the shear layer thickness, will grow linearly as they move downstream due to entrainment of external flow and pairing process and swallowing of existing vortices in the shear layer. It is observed experimentally that subsequent to initial pairing of vortices, a three-dimensional breakdown into turbulence takes place. Thus, imposed on the quasi two-dimensional coherent eddies are the three-dimensional small scale eddies. As in the case of wall-bounded turbulent flows, a significant fraction of turbulent kinetic energy and Reynolds stresses are due to large scale eddies.

20.8 Numerical Techniques for Turbulent Flows

The conservation equations in differential form referred to as the Navier-Stokes equations can be solved numerically to predict transition and evolution of turbulence. In fact the Navier-Stokes equations can be numerically solved for any turbulent flow and it is common to refer to the approach as Direct Numerical Simulation (DNS). This sounds too good to be true and in fact it is at this time. Eventually, as more powerful computers capable of massive computations and having capacity of huge storage are developed, this dream will become a reality. When this will occur is an open question, perhaps within the next 50~100 years! Let's be optimistic.

Several issues must be considered and addressed if one is to use DNS for turbulent flows. First, all scales of turbulence from smallest to largest must be accommodated. This is where limitations to DNS are imposed by present day computers. To identify the problem, recall that the smallest scale of turbulence to consider is the Kolmogorov length scale η, and one can consider the boundary layer thickness δ as the largest scale. To adequately resolve small-scale eddies whose size is of the order of Kolmogorov length scale, a minimum of 4 to 6 grid points are required in each direction. It is rather simple to establish an estimate of the number of grid points required for a DNS of turbulent flows. For example, consider a turbulent boundary layer where the largest length scale is the boundary layer thickness δ. For a three-dimensional computation, the number of grid points with a domain of $\delta \times \delta \times \delta$ would be approximately

$$N = \left(6\frac{\delta}{\eta}\right)^3 = 216\left(\frac{\delta}{\eta}\right)^3$$

where 6 grid points in each direction are used to resolve small scale eddies. Substitution of $\eta = (\nu^3/\epsilon)^{1/4}$ and $\epsilon = U_0^3/\delta$ yields

$$\left(\frac{\delta}{\eta}\right)^3 = \left[\delta\left(\frac{\epsilon}{\nu^3}\right)^{1/4}\right]^3 = \left[\delta\left(\frac{U_0^3\delta^3}{\delta^4\nu^3}\right)^{1/4}\right]^3 = \left(\frac{U_0\delta}{\nu}\right)^{9/4} = (Re_\delta)^{9/4}$$

where

$$Re_\delta = \frac{U_0\delta}{\nu}$$

Thus N is proportional to $(Re_\delta)^{9/4}$ for the three-dimensional DNS of turbulent boundary layer. Simple estimate of the number of grid points for $Re_\delta = 5000$

would be around 4.5×10^{10} grid points. Observe that this is the number of grid points required for a cubical domain of $\delta \times \delta \times \delta$. However, even for a simple problem of flat plate or a channel flow, the domain of solution must be in the order of several δ in each direction, most likely several hundred δ in the streamwise direction. And of course for complex geometries it would be in the order of several million. It is thus seen that the number of required grid points for relatively complex geometries at realistic Reynolds numbers are currently beyond the capability of available computers. Current applications of DNS have been limited to simple geometries at relatively low Reynolds numbers. A review of these applications will be presented in Chapter 23. We can then conclude that, due to the above-mentioned limitations, application of DNS to a majority of problems of interest is ruled out at present. However, this conclusion does not make DNS obsolete, on the contrary, the available results from DNS provide valuable data on the process of transition and development of turbulent flow. These data can be used for development and improvements of turbulence models. Furthermore, DNS provides wide range of data, some of which is very difficult or impossible to measure experimentally. Thus, DNS plays an important role in our understanding and prediction of turbulence and its importance will increase as more powerful computers are developed.

In addition to computer requirements of DNS, several numerical issues must also be addressed. First, the numerical scheme used must be higher order accurate. Recall that typical schemes used in CFD applications are second-order accurate in space. In order to accurately resolve various scales of turbulence an order of accuracy of four or preferably six is required. For this purpose, compact finite difference formulations appear to be very attractive. The compact finite difference formulations will be discussed in Chapter 22.

The second numerical issue is development of accurate boundary conditions. Treatment of boundary conditions will be addressed in Chapter 15.

At this point in time application of DNS for practical problems and its use in design process is ruled out. However, over the last several decades approximate procedures have been developed which allow us to solve turbulent flow fields. The scheme is based on averaging of the fluid properties whereby the high frequency (small-scale) fluctuations are removed. These fluctuating terms are then related to the mean flow properties by relations, which are known as turbulence models. Turbulence models are primarily developed based on experimental data obtained from relatively simple flows under controlled environment. That in turn limits the range of applicability of turbulence models. That is, no single turbulence model is capable of providing accurate solution over a wide range of flow conditions and geometries. Nonetheless, turbulence models are a practical approach for solution of turbulent flows and are used extensively. Discussion of several turbulence models and numerical considerations are provided in Chapter 21.

Finally, a third approach is the large eddy simulation (LES), where large scale structure in the flow are directly simulated whereas small scales are filtered out and are computed by turbulence models called subgrid scale models. Recall that previously it was pointed out that the small scale eddies are more uniform and have more or less common characteristics. Therefore, modeling of small scale turbulence appears more appropriate and the models should apply over wide range of applications. From numerical point of view, since the small-scale turbulence is now modeled, the grid spacing could be much larger than Kolmogorov length scale. This in turn allows applications of LES to larger Reynolds numbers possible. Further discussions of LES and DNS will be provided in Chapter 23.

20.9 Concluding Remarks

- Small-scale eddies appear to be more organized and have similar structure and characteristics in all turbulent flows.
- Large-scale eddies in shear layers are more or less regular and possess coherent structure even though it is possible for them to become irregular. Imposed upon the large scale coherent quasi two-dimensional structures are three-dimensional small scale eddies.
- The growth of large-scale eddies are primarily governed by inertia and pressure effects, whereas viscosity is the primary factor for small scale eddies. In fact, viscosity is the dominant factor limiting the size of small scale eddies.
- Turbulence is an unsteady phenomenon and it is inherently three-dimensional. Eddies overlap in space and larger eddies carry smaller eddies.
- The majority of turbulence production in a boundary layer takes place in the near wall buffer zone where bursting process occurs.
- Kinetic energy contained in large eddies are transferred to smaller eddies and ultimately dissipated into heat at the level of smallest eddies due to molecular viscosity.
- The transfer of energy from larger eddies to smaller and smaller eddies is defined as cascading.
- Kolmogorov length scale is the smallest length scale of turbulence. It is established based on the equilibrium of rate of transfer of energy from the larger eddies to smaller eddies and the dissipation of energy to heat by smallest eddies. The Kolmogrov length scale for a typical flow is about 0.1$\sim$1mm.

- As the flow Reynolds number is increased, the range of eddy sizes becomes wider with the smallest eddies getting smaller.

- An increase in the flow Reynolds number results in longer energy cascade.

- It is perfectly reasonable and logical to solve the governing equations of fluid motion subject to well defined initial and boundary conditions by numerical schemes for turbulent flows. This approach is known as direct numerical simulation (DNS). All relevant scales of turbulence must be accommodated for DNS of turbulent flows. As a result, DNS is extremely computer intensive and for most applications beyond the capabilities of available computers.

- The number of grid points for DNS of turbulent flows scales as $Re^{9/4}$ for three-dimensional problems and as Re^2 for two-dimensional problems. When the smallest scales of the flow are filtered out and the remaining scales are directly computed by the filtered Navier-Stokes equation, the procedure is called *large eddy simulation* (LES). The filtered small scales are determined with the subgrid scale models and thus introduce some degree of uncertainties into the computations.

Chapter 21
Turbulent Flow and Turbulence Models

21.1 Introductory Remarks

Fundamental concepts, basic definitions, conceptual models, and numerical considerations for turbulent flows were established in the previous chapter. Three different techniques for the computation of turbulent flows were identified, and a brief description of each was presented. It was pointed out that, due to large computational time and computational resource requirements for DNS and, to some extent, for LES, these techniques are used more or less for research-oriented applications at this time. For routine computations involving turbulence, the averaged Navier-Stokes equations complemented with turbulence models are utilized. The objectives of this chapter are (a) to review the relevant averaged Navier-Stokes equations and turbulence models, (b) to explore the appropriate numerical schemes for solutions, and (c) to investigate several applications. Discussion of LES and DNS are presented in Chapter 23.

21.2 Fundamental Concepts

In this section, the fundamentals of turbulent flows, some of which were introduced in the previous chapter, are explored and reiterated. This explanation is accomplished by considering the classical turbulent boundary layers.

In many high Reynolds number flows, the effect of viscous forces is dominant in the regions adjacent to the surface. This region is known as the boundary layer. As it was pointed out in the previous chapter, a boundary layer usually starts with well-behaved streamlines where the fluid mixing is at a microscopic level. This type of boundary layer flow is identified as a laminar boundary layer. Due to conditions imposed by the geometry and flow field, such as surface roughness, surface temperature, surface injection, and pressure gradient, the mixing of the fluid increases and takes place at the macroscopic level; streamlines are no longer well-behaved. This type of flow is known as turbulent flow. There is a transitional region between laminar and turbulent boundary layers known as the *transition region*, which was described in Section 20.3. The various boundary layer regions are shown in Figure 21-1.

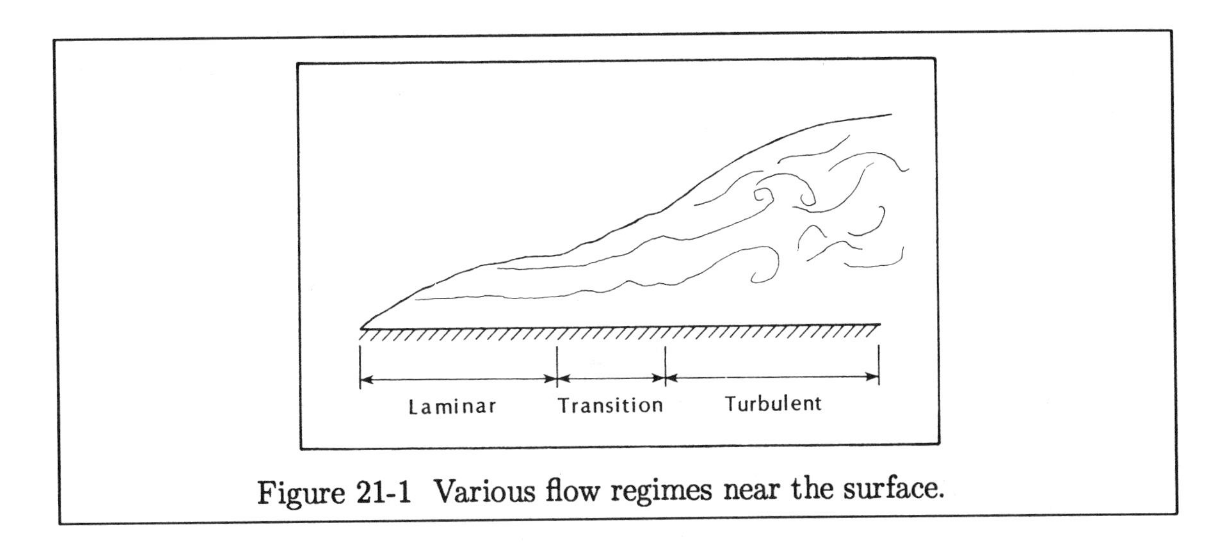

Figure 21-1 Various flow regimes near the surface.

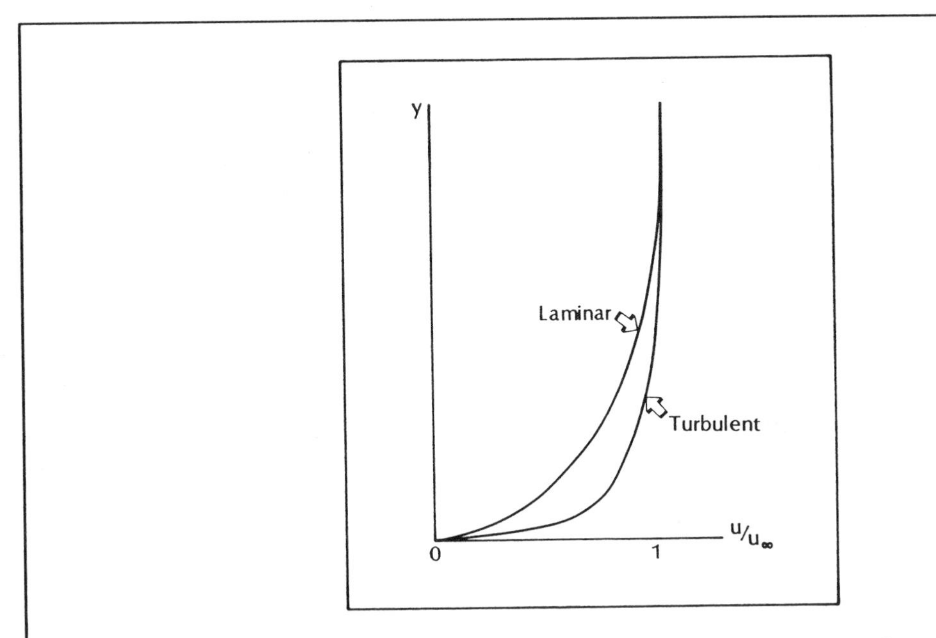

Figure 21-2 Comparison of typical velocity profiles in laminar and turbulent boundary layers.

As a result of heavy mixing of the fluid in a turbulent boundary layer and the associated large momentum flux, the velocity profile becomes fuller in a turbulent boundary layer as compared to a laminar boundary layer; i.e., the velocity gradient near the wall for a turbulent boundary layer is larger than the velocity gradient of a laminar boundary layer. Typical velocity profiles for laminar and turbulent boundary layers are shown in Figure 21-2.

In addition, boundary layer flows can be classified into velocity boundary layers and thermal boundary layers. For most problems, velocity boundary layer thickness and thermal boundary layer thickness are not identical. Typical velocity and thermal boundary layers are shown in Figure 21-3.

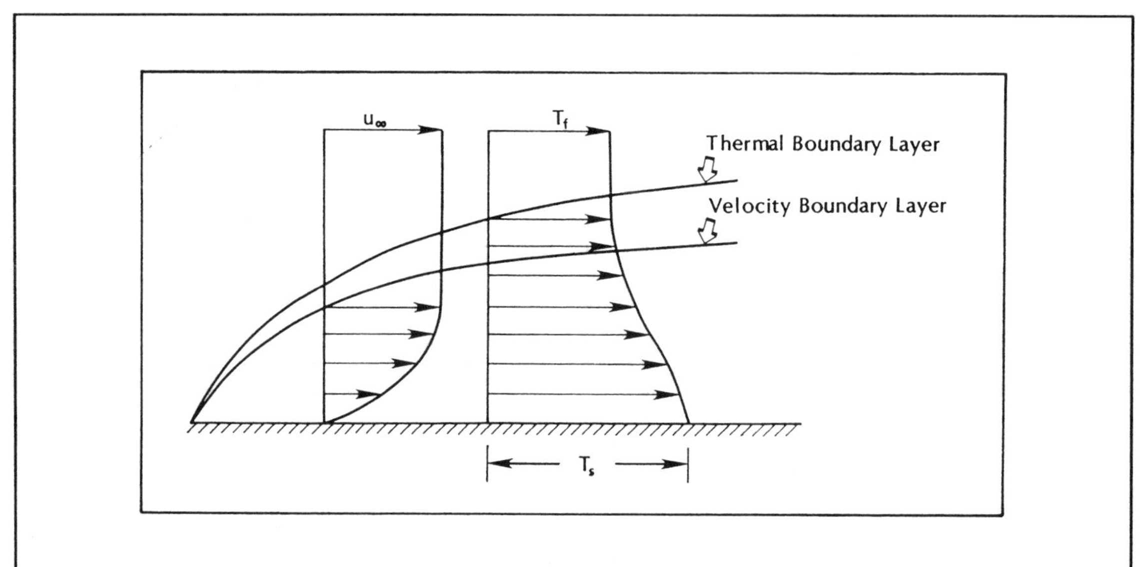

Figure 21-3 A typical comparison of velocity and thermal boundary layers.

Recall that a nondimensional parameter was previously defined as the Prandtl number and expressed as

$$\mathrm{Pr} = \frac{\nu}{\alpha} = \frac{\mu c_p}{k}$$

This parameter represents the ratio of momentum transfer in the flow to that of heat transfer. Therefore, it may represent relative velocity and thermal boundary layer thicknesses. Figure 21-4 illustrates this relation. If δ and δ_t are used to denote the velocity and thermal boundary layer thicknesses, respectively, then,

for $\mathrm{Pr} < 1$, $\delta_t > \delta$

$\mathrm{Pr} = 1$, $\delta_t = \delta$

and for $\mathrm{Pr} > 1$, $\delta_t < \delta$

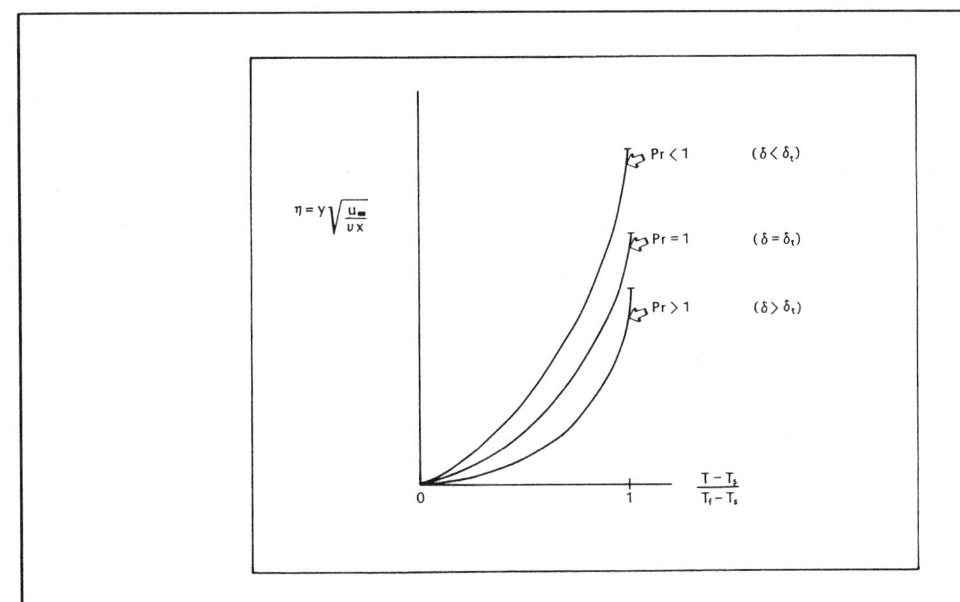

Figure 21-4. The velocity and thermal boundary layer thicknesses for various Prandtl numbers.

21.2.1 Universal Velocity Distribution

A turbulent boundary layer over a smooth flat plate with zero pressure gradient is perhaps one of the simplest type of turbulent flows. However, even this "simple" turbulent flow is in fact very complex, and an analytical solution can not be obtained. It is therefore desirable to develop semi-empirical relations, which can be used to provide approximate solutions for the simple turbulent flows. One such relation is the Universal Velocity Distribution, which is reviewed in this section. This relation is developed based on dimensional analysis and incorporation of experimental data to determine the constants appearing in the expression.

Physical parameters which influence the velocity profile near the wall include density (ρ), viscosity (μ), distance to the wall (y), surface roughness (k), and the velocity gradient at the wall [$(du/dy)_w$], or, equivalently, the shear stress at the wall (τ_w). Thus, a functional relation for the near wall velocity is written as

$$u = F(\rho, \nu, y, k, \tau_w) \tag{21-1}$$

Note that the influence of the outer layer on the near wall velocity profile enters through

$$u^+ = f(y^+, k^+) \tag{21-2}$$

where

$$k^+ = \frac{u_\tau k}{\nu} \tag{21-3}$$

Expression (21-2) is reduced to

$$u^+ = f(y^+) \tag{21-4}$$

for a smooth wall. This relation is known as *the law of the wall*. In terms of the different regions of turbulent boundary layer defined previously in Section 20.2, the law of the wall is valid for the viscous layer, the buffer zone, and the fully turbulent portions of the boundary layer. It should be noted that the total shear stress composed of viscous stress and Reynolds stress is approximately constant within the regime governed by the law of the wall.

The outer region velocity profile can be expressed as

$$u - u_e = G(\rho, y, \delta, u_\tau, dp/dx) \tag{21-5}$$

or in terms of nondimensional parameters as

$$\frac{u - u_e}{u_\tau} = g\left(\frac{y}{\delta}, \frac{\delta}{\rho u_\tau^2}\frac{dp}{dx}\right) \tag{21-6}$$

This relation is known as *the defect law*. Observe that viscosity does not appear in the functional relation (21-5). Recall that as previously discussed in Section 20-20, the effect of viscosity is to dissipate the small eddies due to existence of large velocity gradient. This effect is dominant near the wall where the Reynolds number is relatively low. However, the contribution of small eddies to the Reynolds stress in the high Reynolds number outer region is in fact negligible. As a consequence it is argued that the outer region of the velocity profile is essentially independent of viscosity.

At this point, a relation for the near wall region known as *the law of the wall*, given by the functional form (21-1) or equivalently by (21-2), and a relation for the outer region known as *the defect law*, given by (21-5) or (21-6), has been established based on dimensional analysis and experimental observations. It is then obvious that these relations must merge together smoothly over a finite region between the near wall region and the outer region. This region is known as the log layer or the overlap layer. The length of the log layer is problem dependent, and, in particular, a function of Reynolds number $(u_\tau\delta)/\nu$. Typical range of log layer is $\sim 50 \leq y^+ \leq\sim 400$, or, in terms of y/δ, $\sim 0.01 < y/\delta \leq\sim 0.2$.

Now consider a smooth flat plate with no imposed pressure gradient. Then, from relation (21-4), the law of the wall is

$$\frac{u}{u_\tau} = f\left(\frac{yu_\tau}{\nu}\right) = f\left(\frac{y}{\delta}\frac{u_\tau\delta}{\nu}\right) \tag{21-7}$$

and, from relation (21-6), the defect law is written as

$$\frac{u}{u_\tau} = \frac{u_e}{u_\tau} + g\left(\frac{y}{\delta}\right) \tag{21-8}$$

Since, in the log layer relations, (21-7) and (21-8) must merge together, we set

$$f\left(\frac{u}{\delta}\frac{u_\tau \delta}{\nu}\right) = \frac{u_e}{u_\tau} + g\left(\frac{y}{\delta}\right) \tag{21-9}$$

Functional analysis indicates the validity of relation (21-9) if both f and g have logarithmic form. Thus an equation in the following form is established

$$\frac{u}{u_\tau} = \frac{1}{\kappa}\ell n\, y^+ + B \tag{21-10}$$

where B and κ (von Karman Constant) are nondimensional constants determined from experimental data, with typical values of $\kappa = 0.4 \sim 0.41$ and $B = 5.0 \sim 5.5$. Thus, Equation (21-10) is written as

$$u^+ = 2.5\ell n\, y^+ + 5.5 \tag{21-11}$$

The equation for the outer layer can be expressed as

$$\frac{u - u_e}{u_\tau} = \frac{1}{\kappa}\ell n\frac{y}{\delta} - A \tag{21-12}$$

where

$$A \cong 2.35$$

It is important to emphasize that both relations given by (21-11) and (21-12) are valid only for incompressible flows over smooth flat plates with zero pressure gradient.

Now consider the velocity profile very near the wall within the viscous sublayer, that is, $y^+ < 2 \sim 8$. This is a region where viscous shear dominates and the velocity profile can be approximated to be linear. Therefore,

$$\tau_w = \mu\frac{\partial u}{\partial y} \cong \mu\frac{u}{y}$$

or

$$\frac{\tau_w}{\rho} = \frac{\mu}{\rho}\frac{u}{y} = \nu\frac{u}{y}$$

or

$$u_\tau^2 = \nu\frac{u}{y}$$

from which

$$\frac{u}{u_\tau} = \frac{u_\tau y}{\nu}$$

or

$$u^+ = y^+ \tag{21-13}$$

The velocity profile in the sublayer region given by (21-13) must smoothly merge with the velocity profile for the log layer given by (21-11). A relation satisfying this requirement is established using experimental data as

$$u^+ = 5.0\,\ell n\, y^+ - 3.05 \tag{21-14}$$

which is valid for the buffer zone, that is, $2 \sim 8 < y^+ < 2 \sim 50$.

Now that several relations which essentially form the law of the wall have been established, Figure 20-3 is repeated as Figure 21-5, where the equation for each region is identified. The compressibility law of the wall for the log layer can be derived for most turbulence models. The main assumption used in the derivation is based on the fact that the convection effects are small within the log layer, and therefore they are ignored in the equations of fluid motion and the turbulence models are considered. The mathematical details are omitted here, however the details can be found in References (21.1, 21.2). The resulting expressions from the $k-\epsilon$ model and the $k-\omega$ model are, respectively,

$$u^+ = \frac{1}{\kappa_\epsilon}\,\ell n\, y^+ + B \tag{21-15}$$

where

$$B_\epsilon = B + \frac{1}{\kappa_\epsilon}\ell n \left(\frac{\bar{\rho}}{\bar{\rho}_w}\right)^{5/4} \tag{21-16}$$

and

$$u^+ = \frac{1}{\kappa_\omega}\,\ell n\, y^+ + B \tag{21-17}$$

where

$$B_\omega = B + \frac{1}{\kappa_\omega}\,\ell n \left(\frac{\bar{\rho}}{\bar{\rho}_\omega}\right)^{1/4} \tag{21-18}$$

and $\bar{\rho}_w$ is the density at the wall. The constants κ_ϵ and κ_ω are known as the *effective von Karman constants.* In fact, κ_ϵ and κ_ω are not constants, but are variables for which expressions can be developed, as in Reference (21.1). However, the variation in the values of κ_ϵ and κ_ω is typically small, in the range of about 0.2 percent for κ_ω and 5 percent for κ_ϵ when the Mach number range is up to about 5. Therefore, for simplicity, one may select $\kappa_\omega = \kappa_\epsilon = \kappa = 0.41$.

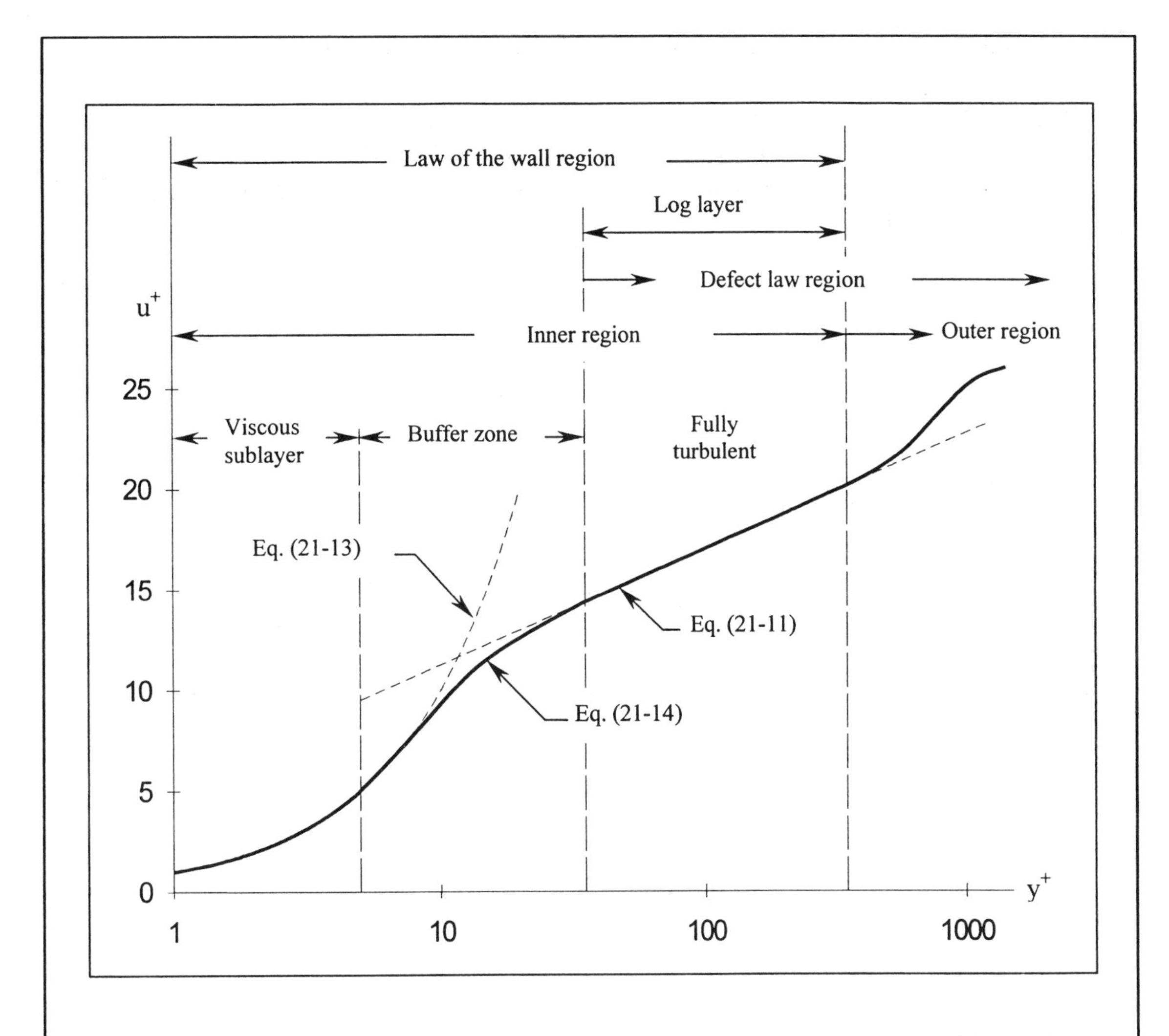

Figure 21-5. Nondimensional velocity profile for an incompressible turbulent flow over a flat plate and identification of different regions within the turbulent boundary layer.

21.3 Modifications of the Equations of Fluid Motion

In order to include and account for the effect of turbulence in a flow field, the equations of fluid motion are modified and amended by turbulence models. There are two approaches to reformulate the Navier-Stokes equations for this purpose. In both approaches, an averaging process is used. The resulting equations are known as the *Reynolds Averaged Navier-Stokes equations* (RANS) and the *Favre Averaged Navier-Stokes equations* (FANS). The governing equations and assumptions for each approach are presented in the section (21.3.1) for RANS and in the section (21.8) for FANS.

21.3.1 Reynolds Averaged Navier-Stokes Equations

Traditionally this modification is accomplished by representing the instantaneous flow quantities by the sum of a mean value (denoted by a bar over the variable) and a time-dependent fluctuating value (denoted by a prime). Mathematically, it is expressed as

$$f = \overline{f} + f' \tag{21-19}$$

where

$$\overline{f} = \frac{1}{\Delta t} \int_{t_o}^{t_o+\Delta t} f dt \tag{21-20}$$

These quantities are shown graphically in Figure 21-6. The time averaging of a fluctuating quantity over a time interval Δt results in

$$\overline{f'} = \frac{1}{\Delta t} \int_{t_o}^{t_o+\Delta t} f' dt = 0 \tag{21-21}$$

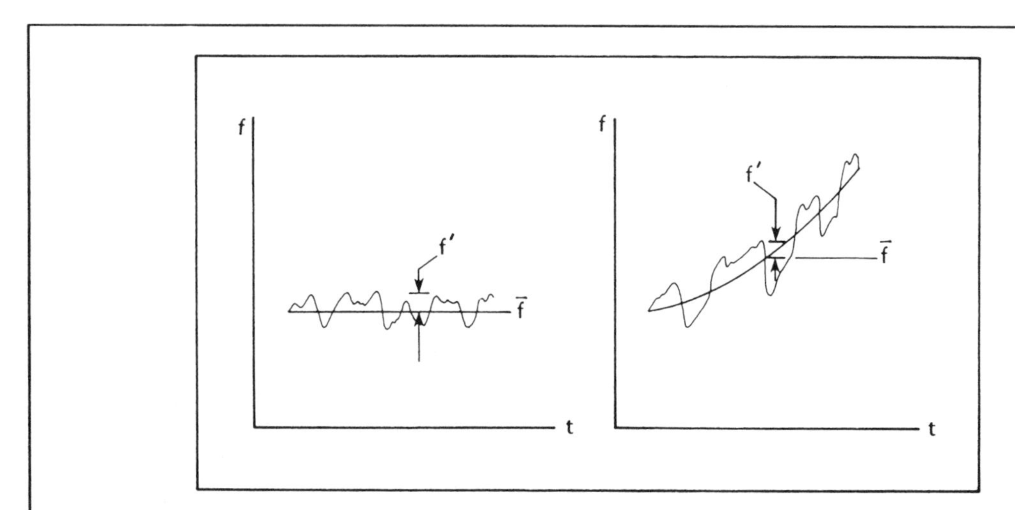

Figure 21-6. Illustration of the average and fluctuating quantities for steady and unsteady flows.

The time interval, Δt, used in the definitions above must be larger than the period of fluctuating quantities but smaller than the time interval associated with the unsteady flow. Thus the time interval is problem dependent, i.e., depends on the geometry and physics of the flow field being investigated. Note that for a steady flow the time-averaged mean value is constant while for an unsteady flow it is a function of time.

In order to carry out the mathematical details, the following averaging rules are applied:

$$\overline{\overline{f}} = \overline{f} \tag{21-22a}$$

$$\overline{f+g} = \overline{f} + \overline{g} \tag{21-22b}$$

$$\overline{\overline{f}g} = \overline{f}\,\overline{g} \tag{21-22c}$$

$$\overline{\frac{\partial f}{\partial x}} = \frac{\partial \overline{f}}{\partial x} \tag{21-22d}$$

$$\overline{f'^2} \neq 0 \tag{21-22e}$$

$$\overline{f'g'} \neq 0 \tag{21-22f}$$

$$\overline{(\overline{f} + f')^2} = \overline{f}^2 + \overline{f'^2} \tag{21-22g}$$

It is important to recognize that the averaged values of the products of fluctuating terms represented by $\overline{f'g'}$ are not, in general, zero. Indeed, these quantities—in particular, those involving the velocity fluctuations—are critically important and represent the cornerstone of turbulent flow effects.

The conservation of mass for a two-dimensional flow is given by the first component of Equation (11-192), which can be expressed in dimensional form as

$$\frac{\partial \rho}{\partial t} + \frac{\partial}{\partial x}(\rho u) + \frac{\partial}{\partial y}(\rho v) + \frac{\alpha}{y}(\rho v) = 0$$

The density and velocity components are replaced by relation (21-19) to provide

$$\frac{\partial}{\partial t}(\overline{\rho} + \rho') + \frac{\partial}{\partial x}\left[(\overline{\rho} + \rho')\,(\overline{u} + u')\right] + \frac{\partial}{\partial y}\left[(\overline{\rho} + \rho')\,(\overline{v} + v')\right]$$

$$+ \frac{\alpha}{y}\left[(\overline{\rho} + \rho')\,(\overline{v} + v')\right] = 0$$

Once this equation is expanded, it is time averaged according to rules set by (21-22) to provide

$$\frac{\partial \overline{\rho}}{\partial t} + \frac{\partial}{\partial x}\left(\overline{\rho}\,\overline{u} + \overline{\rho' u'}\right) + \frac{\partial}{\partial y}\left(\overline{\rho}\,\overline{v} + \overline{\rho' v'}\right) + \frac{\alpha}{y}\left(\overline{\rho}\,\overline{v} + \overline{\rho' v'}\right) = 0 \tag{21-23}$$

The mathematical manipulation for the momentum equation is considered for the x-component, and, subsequently, the results are extended to the y-component. The x-component of the momentum equation in dimensional form is given by the second component of (11-192) as

$$\frac{\partial}{\partial t}(\rho u) + \frac{\partial}{\partial x}(\rho u^2 + p) + \frac{\partial}{\partial y}(\rho u v) + \frac{\alpha}{y}(\rho u v) = \frac{\partial}{\partial x}\left[\mu\left(\frac{4}{3}\frac{\partial u}{\partial x} - \frac{2}{3}\frac{\partial v}{\partial y}\right)\right]$$

$$+ \frac{\partial}{\partial y}\left[\mu\left(\frac{\partial u}{\partial y} + \frac{\partial v}{\partial x}\right)\right] + \frac{\alpha}{y}\left[\mu\left(\frac{\partial u}{\partial y} + \frac{\partial v}{\partial x}\right) - \frac{2}{3}y\frac{\partial}{\partial x}\left(\mu\frac{v}{y}\right)\right]$$

The instantaneous values are replaced by time-averaged mean and fluctuating values to provide

$$\frac{\partial}{\partial t}[(\overline{\rho}+\rho')\ (\overline{u}+u')]+\frac{\partial}{\partial x}\left[(\overline{\rho}+\rho')\ (\overline{u}+u')^2+(\overline{p}+p')\right]$$

$$+\frac{\partial}{\partial y}[(\overline{\rho}+\rho')\ (\overline{u}+u')\ (\overline{v}+v')]+\frac{\alpha}{y}[(\overline{\rho}+\rho')\ (\overline{u}+u')\ (\overline{v}+v')]=$$

$$\frac{\partial}{\partial x}\left\{\mu\left[\frac{4}{3}\frac{\partial}{\partial x}(\overline{u}+u')-\frac{2}{3}\frac{\partial}{\partial y}(\overline{v}+v')\right]\right\}+\frac{\partial}{\partial y}\left\{\mu\left[\frac{\partial}{\partial y}(\overline{u}+u')+\frac{\partial}{\partial x}(\overline{v}+v')\right]\right\}$$

$$+\frac{\alpha}{y}\left\{\mu\left[\frac{\partial}{\partial y}(\overline{u}+u')+\frac{\partial}{\partial x}(\overline{v}+v')\right]-\frac{2}{3}y\frac{\partial}{\partial x}\left[\frac{\mu}{y}(\overline{v}+v')\right]\right\}$$

Since the fluctuating component of viscosity is usually small, it has not been included in this equation. Now the entire equation is time averaged and subsequently rearranged as

$$\frac{\partial}{\partial t}\left(\overline{\rho}\,\overline{u}+\overline{\rho' u'}\right)+\frac{\partial}{\partial x}\left(\overline{\rho}\,\overline{u}^2+\overline{p}+\overline{\rho u'^2}+\overline{\rho' u'^2}+2\overline{u}\overline{\rho' u'}\right)$$

$$\frac{\partial}{\partial y}\left(\overline{\rho}\,\overline{u}\,\overline{v}+\overline{\rho u'v'}+\overline{u}\overline{\rho' v'}+\overline{v}\overline{\rho' u'}+\overline{\rho' u'v'}\right)+\frac{\alpha}{y}\left(\overline{\rho}\,\overline{u}\,\overline{v}+\overline{\rho u'v'}+\overline{u}\overline{\rho' v'}+\overline{v}\overline{\rho' u'}+\overline{\rho' u'v'}\right)=$$

$$\frac{\partial}{\partial x}\left[\mu\left(\frac{4}{3}\frac{\partial\overline{u}}{\partial x}-\frac{2}{3}\frac{\partial\overline{v}}{\partial y}\right)\right]+\frac{\partial}{\partial y}\left[\mu\left(\frac{\partial\overline{u}}{\partial y}+\frac{\partial\overline{v}}{\partial x}\right)\right]+\frac{\alpha}{y}\left[\mu\left(\frac{\partial\overline{u}}{\partial y}+\frac{\partial\overline{v}}{\partial x}\right)-\frac{2}{3}y\frac{\partial}{\partial x}\left(\mu\frac{\overline{v}}{y}\right)\right]$$

Writing this equation in terms of shear stresses, one obtains

$$\frac{\partial}{\partial t}\left(\overline{\rho}\,\overline{u}+\overline{\rho' u'}\right)+\frac{\partial}{\partial x}\left(\overline{\rho}\,\overline{u}^2+\overline{p}+\overline{u}\overline{\rho' u'}\right)+\frac{\partial}{\partial y}\left(\overline{\rho}\,\overline{u}\,\overline{v}+\overline{u}\overline{\rho' v'}\right)+\frac{\alpha}{y}\left(\overline{\rho}\,\overline{u}\,\overline{v}+\overline{u}\overline{\rho' v'}\right)=$$

$$\frac{\partial}{\partial x}\left(\tau_{xxp}-\overline{\rho u'^2}-\overline{\rho' u'^2}-\overline{u}\overline{\rho' u'}\right)+\frac{\partial}{\partial y}\left(\tau_{xy}-\overline{\rho u'v'}-\overline{\rho' u'v'}-\overline{v}\overline{\rho' u'}\right)+$$

$$+\frac{\alpha}{y}\left[\tau_{xy}-\frac{2}{3}y\frac{\partial}{\partial x}\left(\mu\frac{\overline{v}}{y}\right)-\overline{\rho u'v'}-\overline{v}\overline{\rho' u'}-\overline{\rho' u'v'}\right] \tag{21-24}$$

Similarly, the y-component of the momentum equation becomes

$$\frac{\partial}{\partial t}\left(\overline{\rho}\,\overline{v}+\overline{\rho' v'}\right)+\frac{\partial}{\partial x}\left(\overline{\rho}\,\overline{u}\,\overline{v}+\overline{v}\overline{\rho' v'}\right)+\frac{\partial}{\partial y}\left(\overline{\rho}\,\overline{u}^2+\overline{p}+\overline{v}\overline{\rho' v'}\right)$$

$$+\frac{\alpha}{y}\left(\overline{\rho}\,\overline{v}^2+\overline{v}\overline{\rho' v'}\right)=\frac{\partial}{\partial x}\left(\tau_{xy}-\overline{\rho u'v'}-\overline{u}\overline{\rho' v'}-\overline{\rho' u'v'}\right)$$

$$+\frac{\partial}{\partial y}\left(\overline{\tau}_{yyp}-\overline{\rho}\overline{v'^2}-\overline{\rho' v'^2}-\overline{v}\overline{\rho' v'}\right)+\frac{\alpha}{y}\left[\overline{\tau}_{yyp}-\overline{\tau}_{\theta\theta}\right.$$

$$\left.-\frac{2}{3}\frac{\mu}{y}\overline{v}-\frac{2}{3}y\frac{\partial}{\partial y}\left(\mu\frac{\overline{v}}{y}\right)-\overline{\rho}\overline{v'^2}-\overline{\rho' v'^2}-\overline{v}\overline{\rho' v'}\right] \tag{21-25}$$

The energy equation given by the fourth component of (11-192) in terms of the total energy per unit mass is expressed as

$$\frac{\partial}{\partial t}(\rho e_t)+\frac{\partial}{\partial x}\left[(\rho e_t+p)\,u\right]+\frac{\partial}{\partial y}\left[(\rho e_t+p)\,v\right]+\frac{\alpha}{y}\left[(\rho e_t+p)\,v\right]=$$

$$+\frac{\partial}{\partial x}\left[u\tau_{xxp}+v\tau_{xy}-q_x\right]+\frac{\partial}{\partial y}\left[u\tau_{xy}+v\tau_{yyp}-q_y\right]$$

$$+\frac{\alpha}{y}\left[u\tau_{xy}+v\tau_{yyp}-q_y-\frac{2}{3}\mu\frac{v^2}{y}-y\frac{\partial}{\partial y}\left(\frac{2}{3}\mu\frac{v^2}{y}\right)-y\frac{\partial}{\partial x}\left(\frac{2}{3}\mu\frac{uv}{y}\right)\right]$$

Note that the total energy is assumed to be composed of internal energy and kinetic energy. The mathematical procedure used to include turbulence is similar to previous manipulations. The final form of the energy equation becomes

$$\frac{\partial}{\partial t}\left[\overline{\rho}\,\overline{e}_t+\frac{1}{2}\overline{\rho}\left(\overline{u'^2}+\overline{v'^2}\right)+\overline{\rho' e'}+\frac{1}{2}\left(\overline{\rho' u'^2}+\overline{\rho' v'^2}\right)+\overline{u}\overline{\rho' u'}+\overline{v}\overline{\rho' v'}\right]$$

$$+\frac{\partial}{\partial x}\left[\overline{u}\left(\overline{\rho}\,\overline{e}_t+\overline{p}\right)+\frac{1}{2}\overline{\rho}\,\overline{u}\left(\overline{u'^2}+\overline{v'^2}\right)+\overline{u}\left(\overline{\rho' e'}+\overline{u}\overline{\rho' u'}+\overline{v}\overline{\rho' v'}\right)\right.$$

$$+\frac{1}{2}\overline{u}\left(\overline{\rho' u'^2}+\overline{\rho' v'^2}\right)+\overline{\rho}\left(\overline{e' u'}+\frac{1}{2}\overline{u'^3}+\frac{1}{2}\overline{u' v'^2}\right)+\overline{\rho}\left(\overline{u}\overline{u'^2}+\overline{v}\overline{u' v'}\right)$$

$$\left.+\overline{e}_t\overline{\rho' u'}+\overline{\rho' e' u'}+\frac{1}{2}\left(\overline{\rho' u'^3}+2\overline{u}\overline{\rho' u'^2}+\overline{\rho' u' v'^2}+2\overline{v}\overline{\rho' u' v'}\right)+\overline{p' u'}\right]$$

$$+\frac{\partial}{\partial y}\left[\overline{v}\left(\overline{\rho}\,\overline{e}_t+\overline{p}\right)+\frac{1}{2}\overline{\rho}\,\overline{v}\left(\overline{u'^2}+\overline{v'^2}\right)+\overline{v}\left(\overline{\rho' e'}+\overline{u}\overline{\rho' u'}+\overline{v}\overline{\rho' v'}\right)+\right.$$

$$+\frac{1}{2}\overline{v}\left(\overline{\rho' u'^2}+\overline{\rho' v'^2}\right)+\overline{\rho}\left(\overline{e' v'}+\frac{1}{2}\overline{u'^2 v'}+\frac{1}{2}\overline{v'^3}\right)+\overline{\rho}\left(\overline{u}\overline{u' v'}+\overline{v}\overline{v'^2}\right)$$

$$\left.+\overline{e}_t\overline{\rho' v'}+\overline{\rho' e' v'}+\frac{1}{2}\left(\overline{\rho' v' u'^2}+2\overline{u}\overline{\rho' u' v'}+\overline{\rho' v'^3}+2\overline{v}\overline{\rho' v'^2}\right)+\overline{p' v'}\right]$$

$$+\frac{\alpha}{y}\left[\overline{v}\left(\overline{\rho}\,\overline{e}_t+\overline{p}\right)+\frac{1}{2}\overline{\rho}\,\overline{v}\left(\overline{u'^2}+\overline{v'^2}\right)+\overline{v}\left(\overline{\rho' e'}+\overline{u}\overline{\rho' u'}+\overline{v}\overline{\rho' v'}\right)\right.$$

$$+\frac{1}{2}\overline{v}\left(\overline{\rho' u'^2}+\overline{\rho' v'^2}\right)+\overline{\rho}\left(\overline{e' v'}+\frac{1}{2}\overline{u'^2 v'}+\frac{1}{2}\overline{v'^3}\right)+\overline{\rho}\left(\overline{u}\overline{u' v'}+\overline{v}\overline{v'^2}\right)$$

$$+\overline{e}_t\overline{\rho' v'}+\overline{\rho' e' v'}+\frac{1}{2}\left(\overline{\rho' v' u'^2}+2\overline{u}\,\overline{\rho' u' v'}+\overline{\rho' v'^3}+2\overline{v}\,\overline{\rho' v'^2}\right)+\overline{p' v'}\Big]=$$

$$\frac{\partial}{\partial x}\Big[\overline{u}\,\overline{\tau}_{xxp}+\overline{v}\,\overline{\tau}_{xy}-k\frac{\partial\overline{T}}{\partial x}+\mu\left(\frac{4}{3}\overline{u'\frac{\partial u'}{\partial x}}-\frac{2}{3}\overline{u'\frac{\partial v'}{\partial y}}\right)$$

$$+\mu\left(-\frac{2}{3}\left(\overline{v'\frac{\partial u'}{\partial x}}+\overline{v'\frac{\partial v'}{\partial y}}\right)+\frac{4}{3}\frac{\overline{v'^2}}{y}\right)\Big]$$

$$+\frac{\partial}{\partial y}\Big[\overline{u}\,\overline{\tau}_{xy}+\overline{v}\,\overline{\tau}_{yyp}-k\frac{\partial\overline{T}}{\partial y}+\mu\left(-\frac{2}{3}\left(\overline{u'\frac{\partial u'}{\partial x}}+\overline{u'\frac{\partial v'}{\partial y}}\right)+\frac{4}{3}\frac{1}{y}\overline{u'v'}\right)$$

$$+\mu\left(\frac{4}{3}\overline{v'\frac{\partial v'}{\partial y}}-\frac{2}{3}\overline{v'\frac{\partial u'}{\partial x}}\right)\Big]$$

$$+\frac{\alpha}{y}\Big[\overline{u}\,\overline{\tau}_{xy}+\overline{v}\,\overline{\tau}_{yyp}-k\frac{\partial\overline{T}}{\partial y}+\mu\left(-\frac{2}{3}\left(\overline{u'\frac{\partial u'}{\partial x}}+\overline{u'\frac{\partial v'}{\partial y}}\right)+\frac{4}{3}\frac{1}{y}\overline{u'v'}\right)$$

$$+\mu\left(\frac{4}{3}\overline{v'\frac{\partial v'}{\partial y}}-\frac{2}{3}\overline{v'\frac{\partial u'}{\partial x}}\right)-\frac{2}{3}\frac{\mu}{y}\left(\overline{v}^2\right)-\frac{2}{3}\frac{\mu}{y}\left(\overline{v'^2}\right)$$

$$-y\frac{\partial}{\partial y}\left(\frac{2}{3}\frac{\mu}{y}\left(\overline{v}^2+\overline{v'^2}\right)\right)-y\frac{\partial}{\partial x}\left(\frac{2}{3}\frac{\mu}{y}\left(\overline{u}\,\overline{v}+\overline{u'v'}\right)\right)\Big] \tag{21-26}$$

Now it is obvious that the number of unknowns in the system of equations composed of (21-23) through (21-26) has increased significantly. Before considering additional relations to close the system, some reduction of terms is introduced. It has been shown experimentally that, for flows up to about Mach five, turbulence structure remains unchanged and is similar to that of incompressible flows (the so-called Morkovin hypothesis). Therefore, the fluctuating correlations involving density fluctuations are usually neglected. In addition, triple correlations such as $\overline{v'^3}$ are assumed smaller than the double correlations and are neglected as well. Finally, Equations (21-23) through (21-26) are expressed as

$$\frac{\partial Q}{\partial t}+\frac{\partial Q'}{\partial t}+\frac{\partial E}{\partial x}+\frac{\partial F}{\partial y}+\alpha H=\frac{\partial E_v}{\partial x}+\frac{\partial F_v}{\partial y}+\alpha H_v+\frac{\partial E'}{\partial x}+\frac{\partial F'}{\partial y}+\alpha H' \tag{21-27}$$

where

$$Q=\begin{bmatrix}\overline{\rho}\\ \overline{\rho}\,\overline{u}\\ \overline{\rho}\,\overline{v}\\ \overline{\rho}\,\overline{e}_t\end{bmatrix} \quad (21\text{-}28a) \qquad Q'=\begin{bmatrix}0\\ 0\\ 0\\ \frac{1}{2}\overline{\rho}\left(\overline{u'^2}+\overline{v'^2}\right)\end{bmatrix} \quad (21\text{-}28b)$$

$$E = \begin{bmatrix} \bar{\rho}\,\bar{u} \\ \bar{\rho}\,\bar{u}^2 + \bar{p} \\ \bar{\rho}\,\bar{u}\,\bar{v} \\ (\bar{\rho}\,\bar{e}_t + \bar{p})\,\bar{u} \end{bmatrix} \quad (21\text{-}28c) \qquad F = \begin{bmatrix} \bar{\rho}\,\bar{v} \\ \bar{\rho}\,\bar{u}\,\bar{v} \\ \bar{\rho}\,\bar{v}^2 + \bar{p} \\ (\bar{\rho}\,\bar{e}_t + \bar{p})\,\bar{v} \end{bmatrix} \quad (21\text{-}28d)$$

$$H = \frac{1}{y}\begin{bmatrix} \bar{\rho}\,\bar{v} \\ \bar{\rho}\,\bar{u}\,\bar{v} \\ \bar{\rho}\,\bar{v}^2 \\ (\bar{\rho}\,\bar{e}_t + \bar{p})\,\bar{v} \end{bmatrix} \quad (21\text{-}28e) \qquad E_v = \begin{bmatrix} 0 \\ \bar{\tau}_{xxp} \\ \bar{\tau}_{xy} \\ \bar{u}\,\bar{\tau}_{xxp} + \bar{v}\,\bar{\tau}_{xy} - \bar{q}_x \end{bmatrix} \quad (21\text{-}28f)$$

$$F_v = \begin{bmatrix} 0 \\ \bar{\tau}_{xy} \\ \bar{\tau}_{yyp} \\ \bar{u}\,\bar{\tau}_{xy} + \bar{v}\,\bar{\tau}_{yyp} - q_y \end{bmatrix} \quad (21\text{-}28g)$$

$$H_v = \frac{1}{y}\begin{bmatrix} 0 \\ \bar{\tau}_{xy} - \frac{2}{3}y\frac{\partial}{\partial x}\left(\mu\frac{\bar{v}}{y}\right) \\ \bar{\tau}_{yyp} - \bar{\tau}_{\theta\theta} - \frac{2}{3}\frac{\mu}{y}\bar{v} - \frac{2}{3}y\frac{\partial}{\partial y}\left(\mu\frac{\bar{v}}{y}\right) \\ \bar{u}\,\bar{\tau}_{xy} + \bar{v}\,\bar{\tau}_{yyp} - \bar{q}_y - \frac{2}{3}\frac{\mu}{y}\left(\bar{v}^2\right) - y\frac{\partial}{\partial y}\left(\frac{2}{3}\mu\frac{\bar{v}^2}{y}\right) - y\frac{\partial}{\partial x}\left(\frac{2}{3}\mu\frac{\bar{u}\bar{v}}{y}\right) \end{bmatrix} \quad (21\text{-}28h)$$

$$E' = \begin{bmatrix} 0 \\ -\bar{\rho}\overline{u'^2} \\ -\bar{\rho}\overline{u'v'} \\ -\frac{1}{2}\bar{\rho}\,\bar{u}\left(\overline{u'^2} + \overline{v'^2}\right) - \bar{\rho}\left(\overline{e'u'}\right) - \bar{\rho}\,\bar{u}\left(\overline{u'^2}\right) - \bar{\rho}\,\bar{v}\left(\overline{u'v'}\right) - \overline{p'u'} \\ +\mu\left(\frac{4}{3}\overline{u'\frac{\partial u'}{\partial x}} - \frac{2}{3}\overline{u'\frac{\partial v'}{\partial y}}\right) + \mu\left[-\frac{2}{3}\left(\overline{v'\frac{\partial u'}{\partial x}} + \overline{v'\frac{\partial v'}{\partial y}}\right) + \frac{4}{3}\frac{\overline{v'^2}}{y}\right] \end{bmatrix} \quad (21\text{-}29a)$$

$$F' = \begin{bmatrix} 0 \\ -\bar{\rho}\overline{u'v'} \\ -\bar{\rho}\overline{v'^2} \\ -\frac{1}{2}\bar{\rho}\,\bar{v}\left(\overline{u'^2} + \overline{v'^2}\right) - \bar{\rho}\left(\overline{e'v'}\right) - \bar{\rho}\,\bar{u}\left(\overline{u'v'}\right) - \bar{\rho}\,\bar{v}\left(\overline{v'^2}\right) - \overline{p'v'} \\ +\mu\left[-\frac{2}{3}\left(\overline{u'\frac{\partial u'}{\partial x}} + \overline{u'\frac{\partial v'}{\partial y}}\right) + \frac{4}{3}\frac{1}{y}\overline{u'v'}\right] + \mu\left(\frac{4}{3}\overline{v'\frac{\partial v'}{\partial y}} - \frac{2}{3}\overline{v'\frac{\partial u'}{\partial x}}\right) \end{bmatrix} \quad (21\text{-}29b)$$

$$H' = \frac{1}{y}\begin{bmatrix} 0 \\ -\overline{\rho}\overline{u'v'} \\ -\overline{\rho}\overline{v'^2} \\ -\frac{1}{2}\overline{\rho}\,\overline{v}\left(\overline{u'^2} + \overline{v'^2}\right) - \overline{\rho}\left(\overline{e'v'}\right) - \overline{\rho}\,\overline{u}\left(\overline{u'v'}\right) - \overline{\rho}\,\overline{v}\left(\overline{v'^2}\right) - \overline{p'v'} \\ +\mu\left[-\frac{2}{3}\left(\overline{u'\frac{\partial u'}{\partial x}} + \overline{u'\frac{\partial v'}{\partial y}}\right) + \frac{4}{3}\frac{1}{y}\overline{u'v'}\right] + \mu\left(\frac{4}{3}\overline{v'\frac{\partial v'}{\partial y}} - \frac{2}{3}\overline{v'\frac{\partial u'}{\partial x}}\right) \\ -\frac{2}{3}\frac{\mu}{y}\left(\overline{v'^2}\right) - y\frac{\partial}{\partial y}\left[\frac{2}{3}\frac{\mu}{y}\left(\overline{v'^2}\right)\right] - y\frac{\partial}{\partial x}\left(\frac{2}{3}\frac{\mu}{y}\overline{u'v'}\right) \end{bmatrix} \qquad (21\text{-}29c)$$

Additional unknowns appear in the system of equations given by (21-27) as a result of modification of the original system (11-192) to include the effect of turbulence. It is obvious that the solution of Equation (21-27), which includes turbulence quantities, would be a very challenging task. In fact, in practice, the effect of turbulence in most applications is simulated by the introduction of turbulent viscosity and turbulent conductivity, as will be shown in Section 21.3.1.1. In either case, there are additional unknowns in the Navier-Stokes equations, due to turbulence.

In order to close the system, supplementary relations must be introduced. These relations are known as turbulence models, which in general relate the fluctuating correlations to the mean flow properties by means of empirical constants. Turbulence models vary in degree of sophistication from simple algebraic equations to systems of partial differential equations. Surprisingly enough, in many instances simple algebraic models provide as good a result as sophisticated models, with less computation time. Obviously, the ultimate goal is to simulate and resolve all scales of turbulence directly by the Navier-Stokes equations. However, due to limitations in the capacity of current computers, the direct simulation approach is more or less at the research level, and it has not matured for routine computations as yet. With the advancement in computer technology, however, this goal will be accomplished in the future. Research toward this goal is underway.

21.3.1.1 Turbulent Shear Stress and Heat Flux:

In this section, some of the fundamental concepts introduced by the early investigating pioneers are explored. For the sake of simplicity, the steady incompressible boundary layer equations are considered. The governing equations are expressed as

$$\frac{\partial \overline{u}}{\partial x} + \frac{\partial \overline{v}}{\partial y} = 0 \qquad (21\text{-}30)$$

$$\overline{u}\frac{\partial \overline{u}}{\partial x} + \overline{v}\frac{\partial \overline{u}}{\partial y} = -\frac{1}{\rho}\frac{\partial \overline{p}}{\partial x} + \nu\frac{\partial^2 \overline{u}}{\partial y^2} - \frac{\partial}{\partial y}\left(\overline{u'v'}\right) \qquad (21\text{-}31)$$

$$\overline{u}\frac{\partial \overline{T}}{\partial x} + \overline{v}\frac{\partial \overline{T}}{\partial y} = \frac{\nu}{c_p}\left(\frac{\partial \overline{u}}{\partial y}\right)^2 - \frac{1}{c_p}\left(\overline{u'v'}\right)\frac{\partial \overline{u}}{\partial y} + \alpha\frac{\partial^2 \overline{T}}{\partial y^2} - \frac{\partial}{\partial y}\left(\overline{v'T'}\right) \quad (21\text{-}32)$$

In the equation above, α is the thermal diffusivity.

It can be shown that the macroscopic momentum exchange due to turbulence is represented by $-\rho\overline{u'v'}$, which is referred to as the *turbulent shear stress* (or *Reynolds stress*), denoted herein by τ_t. It is customary to combine the laminar and turbulent shear stress terms in Equation (21-31) as

$$\nu\frac{\partial^2 \overline{u}}{\partial y^2} - \frac{\partial}{\partial y}\left(\overline{u'v'}\right) = \frac{1}{\rho}\frac{\partial}{\partial y}(\tau_l + \tau_t) \quad (21\text{-}33)$$

For a laminar flow under the assumptions stated, one may write

$$\tau_l = \mu\frac{\partial \overline{u}}{\partial y} = \rho\nu\frac{\partial \overline{u}}{\partial y} \quad (21\text{-}34)$$

In order to express the turbulent shear in a similar form, consider the concept of eddy viscosity, where the turbulent shear stress is related to the gradient of the mean flow velocity. This analogy was introduced by Boussinesq and is referred to as the *Boussinesq assumption.* Thus, one writes

$$\tau_t = \mu_t\frac{\partial \overline{u}}{\partial y} = \rho\nu_t\frac{\partial \overline{u}}{\partial y} \quad (21\text{-}35)$$

where μ_t (or ν_t) is known as the turbulent viscosity or eddy viscosity. Hence, Equation (21-31) may be expressed as

$$\overline{u}\frac{\partial \overline{u}}{\partial x} + \overline{v}\frac{\partial \overline{u}}{\partial y} = -\frac{1}{\rho}\frac{\partial \overline{p}}{\partial x} + (\nu + \nu_t)\frac{\partial^2 \overline{u}}{\partial^2 y} = -\frac{1}{\rho}\frac{\partial \overline{p}}{\partial y} + \frac{1}{\rho}(\mu + \mu_t)\frac{\partial^2 \overline{u}}{\partial^2 y} \quad (21\text{-}36)$$

In a similar fashion, turbulent conductivity is defined to combine laminar and turbulent heat fluxes. For a laminar flow, the Fourier heat conduction law is expressed as

$$\left(\frac{q}{A}\right)_l = -k\frac{\partial \overline{T}}{\partial y} = -\rho c_p \alpha\frac{\partial \overline{T}}{\partial y} \quad (21\text{-}37)$$

For turbulent heat flux, the following is written,

$$\left(\frac{q}{A}\right)_t = -k_t\frac{\partial \overline{T}}{\partial y} = -\rho c_p \alpha_t\frac{\partial \overline{T}}{\partial y} \quad (21\text{-}38)$$

where k_t is the turbulent conductivity and α_t is the turbulent diffusivity. Now, the energy equation may be rearranged as

$$\overline{u}\frac{\partial \overline{T}}{\partial x} + \overline{v}\frac{\partial \overline{T}}{\partial y} = \frac{1}{\rho c_p}(\tau_l + \tau_t)\frac{\partial \overline{u}}{\partial y} - \frac{1}{\rho c_p}\frac{\partial}{\partial y}\left[\left(\frac{q}{A}\right)_l + \left(\frac{q}{A}\right)_t\right] \quad (21\text{-}39)$$

or in terms of turbulent parameters,

$$\overline{u}\frac{\partial \overline{T}}{\partial x}+\overline{v}\frac{\partial \overline{T}}{\partial y}=\frac{1}{\rho c_p}(\mu+\mu_t)\left(\frac{\partial \overline{u}}{\partial y}\right)^2+\frac{1}{\rho c_p}(k+k_t)\frac{\partial^2 \overline{T}}{\partial y^2} \tag{21-40}$$

Note that, up to this point, the problem of closure has not been addressed, i.e., two additional unknowns still remain. All that has been accomplished is the rewriting of the equations by introducing the Boussinesq assumption.

It is well known that in a laminar flow the mixing of fluid is at the molecular level, and the viscous stresses and heat fluxes are due to momentum and energy transfer by molecules traveling a distance of mean-free path before collision. A similar concept is extended to turbulent flows, where it is assumed that lumps of fluid travel a finite distance before collision and before losing their identity. The resulting momentum and energy exchange give rise to what was defined as turbulent shear stress and turbulent heat flux. This finite distance is defined as the mixing length, l_m. This concept was introduced by Prandtl and is known as the Prandtl mixing length. The Prandtl hypothesis is expressed as

$$\tau_t=-\rho\overline{u'v'}=\rho l_m^2\left(\frac{\partial \overline{u}}{\partial y}\right)^2 \tag{21-41}$$

Similarly,

$$\left(\frac{q}{A}\right)_t=\rho l_e^2\left(\frac{\partial \overline{T}}{\partial y}\right)^2 \tag{21-42}$$

In terms of turbulent viscosity and turbulent conductivity, the following may be written:

$$\mu_t = \rho l_m^2\left(\frac{\partial \overline{u}}{\partial y}\right) \tag{21-43}$$

$$k_t = \rho c_p l_e^2\left(\frac{\partial \overline{T}}{\partial y}\right) \tag{21-44}$$

The concept of turbulent viscosity and turbulent conductivity can be easily extended to general flow fields governed by the Navier-Stokes equations. For example, for a 2-D problem,

$$\mu_t=\rho l_m^2\left[\left(\frac{\partial \overline{u}}{\partial y}\right)^2+\left(\frac{\partial \overline{v}}{\partial x}\right)^2\right]^{\frac{1}{2}} \tag{21-45}$$

What has been accomplished up to this point is the introduction of new parameters, and the central issue of closure is not complete as yet since one still has two additional unknowns, now in the form of μ_t and k_t, or, equivalently, l_m and l_e.

In order to write expressions for mixing lengths, experimental investigations are heavily relied upon; and l_m and l_e are modeled for the various flow regimes, namely the inner and outer regions.

Before proceeding further, it is beneificial to elaborate on the concept of mixing length l_m and the turbulence length scale l, because both are used in the development of turbulence models. Recall that previously, in Chapter 20, the turbulence length scale was defined as a characteristic length associated with the eddy size in a turbulent flow. The smallest of the length scales was defined as *Kolmogorov length scale* associated with the smallest eddy. It was also established that the size of the length scale within a flow varies substantially from the smallest scale (Kolmogorov length scale) to the largest scale (the size of boundary layer or shear layer thicknesses). Furthermore, length scales are flow dependent and would be different for different flow conditions.

The concept of mixing length proposed by Prandtl was developed based on the analogy of momentum exchange in laminar and turbulent flows. Thus, mixing length is a nonphysical property. The mixing length concept was used by pioneering investigators in the development of turbulence models expressed as algebraic relations. Later on, when one- and two-equation models were developed, the concept of turbulence length scale was incorporated in the expressions for turbulence quantitites. Thus, the mixing length and the turbulence length scale are in fact different quantities, and in general the mixing length would not be equal to the turbulence length scale. However, for a special case where the ratio of production to dissipation remains constant, the length scale would be proportional to the mixing length, that is, $l = \text{constant} \cdot l_m$.

In the introduction of turbulence models to follow, the mixing length concept will be used primarily for the algebraic turbulence models, whereas turbulence length scale is used in the one- and two-equation turbulence models.

21.3.1.2 Flux Vector RANS Formulation:

A simple and practical approach to include the effect of turbulence in the Navier-Stokes equations is simply to replace the molecular viscosity μ in the stress terms by $(\mu + \mu_t)$ and the term (μ/Pr) in the heat conduction term by $(\mu/Pr + \mu_t/Pr_t)$. The parameter Pr_t is called the *turbulent Prandtl number*, and for air it is generally taken to be 0.9 for wall bounded flows and 0.5 for shear layers.

Returning to the nondimensional Navier-Stokes equations in the computational space given by Equation (11-60) and repeated here for convenience, one has

$$\frac{\partial \bar{Q}}{\partial \tau} + \frac{\partial \bar{E}}{\partial \xi} + \frac{\partial \bar{F}}{\partial \eta} + \frac{\partial \bar{G}}{\partial \zeta} = \frac{\partial \bar{E}_v}{\partial \xi} + \frac{\partial \bar{F}_v}{\partial \eta} + \frac{\partial \bar{G}_v}{\partial \zeta} \qquad (21\text{-}46)$$

where the flux vectors are given by (11-62) through (11-67). Now, to include the effect of turbulence, Equation (21-46) is used along with the following definitions for the viscous stress and heat conduction terms.

$$\tau_{xx} = \frac{1}{Re_\infty}(\mu+\mu_t)\left[\frac{4}{3}(\xi_x u_\xi + \eta_x u_\eta + \zeta_x u_\zeta) - \frac{2}{3}(\xi_y v_\xi + \eta_y v_\eta + \zeta_y v_\zeta) - \frac{2}{3}(\xi_z w_\xi + \eta_z w_\eta + \zeta_z w_\zeta)\right] \quad (21\text{-}47)$$

$$\tau_{yy} = \frac{1}{Re_\infty}(\mu+\mu_t)\left[\frac{4}{3}(\xi_y v_\xi + \eta_y v_\eta + \zeta_y v_\zeta) - \frac{2}{3}(\xi_x u_\xi + \eta_x u_\eta + \zeta_x u_\zeta) - \frac{2}{3}(\xi_z w_\xi + \eta_z w_\eta + \zeta_z w_\zeta)\right] \quad (21\text{-}48)$$

$$\tau_{zz} = \frac{1}{Re_\infty}(\mu+\mu_t)\left[\frac{4}{3}(\xi_z w_\xi + \eta_z w_\eta + \zeta_z w_\zeta) - \frac{2}{3}(\xi_x u_\xi + \eta_x u_\eta + \zeta_x u_\zeta) - \frac{2}{3}(\xi_y v_\xi + \eta_y v_\eta + \zeta_y v_\zeta)\right] \quad (21\text{-}49)$$

$$\tau_{xy} = \tau_{yx} = \frac{1}{Re_\infty}(\mu+\mu_t)\left(\xi_y u_\xi + \eta_y u_\eta + \zeta_y u_\zeta + \xi_x v_\xi + \eta_x v_\eta + \zeta_x v_\zeta\right) \quad (21\text{-}50)$$

$$\tau_{xz} = \tau_{zx} = \frac{1}{Re_\infty}(\mu+\mu_t)\left(\xi_z u_\xi + \eta_z u_\eta + \zeta_z u_\zeta + \xi_x w_\xi + \eta_x w_\eta + \zeta_x w_\zeta\right) \quad (21\text{-}51)$$

$$\tau_{yz} = \tau_{zy} = \frac{1}{Re_\infty}(\mu+\mu_t)\left(\xi_z v_\xi + \eta_z v_\eta + \zeta_z v_\zeta + \xi_y w_\xi + \eta_y w_\eta + \zeta_y w_\zeta\right) \quad (21\text{-}52)$$

$$q_x = -\frac{1}{Re_\infty(\gamma-1)M_\infty^2}\left(\frac{\mu}{Pr}+\frac{\mu_t}{Pr_t}\right)\left(\xi_x T_\xi + \eta_x T_\eta + \zeta_x T_\zeta\right) \quad (21\text{-}53)$$

$$q_y = -\frac{1}{Re_\infty(\gamma-1)M_\infty^2}\left(\frac{\mu}{Pr}+\frac{\mu_t}{Pr_t}\right)\left(\xi_y T_\xi + \eta_y T_\eta + \zeta_y T_\zeta\right) \quad (21\text{-}54)$$

$$q_z = -\frac{1}{Re_\infty(\gamma-1)M_\infty^2}\left(\frac{\mu}{Pr}+\frac{\mu_t}{Pr_t}\right)\left(\xi_z T_\xi + \eta_z T_\eta + \zeta_z T_\zeta\right) \quad (21\text{-}55)$$

where the value of μ_t is provided by a turbulence model.

21.4 Turbulence Models

A turbulence model is a semi-empirical equation relating the fluctuating correlation to mean flow variables with various constants provided from experimental investigations. When this equation is expressed as an algebraic equation, it is referred to as the *zero-equation model*. When partial differential equations are used, they are referred to as *one-equation* or *two-equation models*, depending on the number of PDEs utilized. Some models employ ordinary differential equations, in which case they are referred to as half-equation models. Finally, it is possible to write a partial differential equation directly for each of the turbulence correlations such as $\overline{u'v'}$, $\overline{u'^2}$, etc., in which case they compose a system of PDEs known as the Reynolds stress equations.

In the following sections, some commonly used zero-equation, one-equation, and two-equation turbulence models, as a sample of the wide range of models, are investigated. Several recently published texts, survey papers, and comparative analysis papers on turbulence models are excellent references for those interested in various aspects of turbulence models, [21-1] through [21-8].

21.4.1 Zero-Equation Turbulence Models

The zero-equation models are equations wherein the turbulent fluctuating correlations are related to the mean flow field quantities by algebraic relations. The underlying assumption in zero-equation models is that the local rate of production of turbulence and the rate of dissipation of turbulence are approximately equal. Furthermore, they do not include the convection of turbulence. Obviously, this is contrary to the physics of most flow fields, since the past history of the flow must be accounted for. Nevertheless, these models are mathematically simple and their incorporation into a numerical code can be accomplished with relative ease.

Generally, most models employ an inner region/outer region formulation to represent mixing length (the subscript m will be dropped from l_m in the following). A commonly used model utilizes an exponential function (van Driest damping function) for the inner region, whereas the outer region is proportional to the boundary layer thickness. Mathematically they are expressed as

$$l_i = \kappa(1 - e^{-y^+/A^+})y \tag{21-56}$$

and

$$l_o = C_o\delta \tag{21-57}$$

where κ is the von Karman constant (~ 0.41), and A^+ is a parameter which depends on the streamwise pressure gradient. For a zero-pressure gradient flow, it has a value of 26. The constant C_o in Equation (21-57) is usually assigned a value of $0.08 \sim 0.09$, whereas δ is the boundary layer thickness.

Another formulation commonly used for the outer region turbulent viscosity is the Cebeci/Smith model expressed as

$$\nu_t = \alpha \bar{u}_e \delta^* \tag{21-58}$$

where α is usually assigned a value of 0.0168 for flows where Re_θ (momentum thickness Reynolds number) is greater than 5000 and δ^* is the kinetic displacement thickness defined as

$$\delta^* = \int_o^\infty \left(1 - \frac{u}{u_e}\right) dy \qquad \textit{for an incompressible flow, and} \qquad (21\text{-}59)$$

$$\delta^* = \int_o^\infty \left(1 - \frac{\rho u}{\rho_e u_e}\right) dy \qquad \textit{for a compressible flow} \qquad (21\text{-}60)$$

Recall that the Reynolds number based on momentum thickness is defined as

$$Re_\theta = \frac{\rho_e u_e \theta}{\mu_e} \qquad (21\text{-}61)$$

where the momentum thickness θ is

$$\theta = \int_o^\infty \frac{u}{u_e}\left(1 - \frac{u}{u_e}\right) dy \qquad \textit{for an incompressible flow, and} \qquad (21\text{-}62)$$

$$\theta = \int_o^\infty \frac{\rho u}{\rho_e u_e}\left(1 - \frac{u}{u_e}\right) dy \qquad \textit{for a compressible flow} \qquad (21\text{-}63)$$

The algebraic model described above requires information regarding the boundary layer thickness and flow properties at the boundary layer edge. When the Navier-Stokes equations are being solved, it may be a difficult task to determine the boundary layer thickness and the required properties at the edge. That is especially the case when flow separation exists within the domain. However, when it is necessary to determine the extent of the viscous region within the domain, the total enthalpy is usually used.

A turbulence model which is not written in terms of the boundary layer quantities was introduced by Baldwin and Lomax [21-9]. The inner region is approximated by

$$\mu_t = \rho l^2 |\omega| \qquad (21\text{-}64)$$

where ω is the vorticity defined as

$$\omega = \frac{\partial v}{\partial x} - \frac{\partial u}{\partial y} \qquad (21\text{-}65)$$

and l is given by (21-56). The nondimensional space coordinate y^+ can be written as

$$y^+ = u_\tau \frac{y}{\nu_w} = \sqrt{\frac{|\tau_w|}{\rho_w}} \frac{\rho_w}{\mu_w} y$$

or

$$y^+ = \sqrt{\rho_w |\tau_w|} \frac{y}{\mu_w}$$

where the subscript w indicates the wall quantities. The outer region is approximated by

$$\mu_t = \alpha \bar{\rho} C_{CP} F_{\text{wake}} F_{\text{Kleb}} \tag{21-66}$$

where α is assigned a value of 0.0168 (as in the Cebeci/Smith model) and

$$F_{\text{wake}} = \min \left[y_{\max} G_{\max} \; , \quad C_{\text{wake}} y_{\max} \frac{(\Delta V)^2}{G_{\max}} \right] \tag{21-67}$$

A typical value for C_{wake} is 1.0. In Equation (21-67), the following definitions are employed,

$$G_{\max} = \max \left(\frac{l}{\kappa} |\omega| \right) \tag{21-68}$$

where κ is the von Karman constant, and l, the mixing length, is determined by the van Driest function given by (21-56). The difference between the absolute values of the maximum and minimum velocities within the viscous region is denoted by ΔV. For wall bounded flows, the minimum velocity occurs at the surface where the velocity is zero, then

$$\Delta V = (u^2 + v^2)^{1/2}_{\max}$$

For shear layer flows, ΔV is defined as the difference between the maximum velocity and the velocity at the $y_{\max}$ location, that is,

$$\Delta V = (u^2 + v^2)^{1/2}_{\max} - (u^2 + v^2)^{1/2}_{y_{\max}}$$

F_{Kleb} is the intermittency factor defined as

$$F_{\text{Kleb}} = \left[1 + 5.5 \left(\frac{C_{\text{Kleb}} y}{y_{\max}} \right)^6 \right]^{-1} \tag{21-69}$$

and $y_{\max}$ is the y location where $G_{\max}$ occurs.

The typical values for the Klebanoff constant C_{Kleb} and C_{CP} are respectively 0.3 and 1.6 for zero or mild pressure gradient. However, to include the influence of pressure gradient, the following expressions as suggested in Reference [21-10] may be used:

$$C_{\text{Kleb}} = \frac{2}{3} - \frac{0.01312}{0.1724 - \beta^*} \tag{21-70}$$

where

$$\beta^* = \frac{y_{\max}}{u_\tau} \frac{\partial V}{\partial x}$$

and u_τ is the friction velocity. The velocity gradient in β^* is calculated outside the viscous region. Once the Klebanoff constant is evaluated, C_{CP} is determined from

$$C_{CP} = \frac{3 - 4C_{\text{Kleb}}}{2C_{\text{Kleb}}(2 - 3C_{\text{Kleb}} + C_{\text{Kleb}}^3)} \tag{21-71}$$

Finally, the turbulent viscosity distribution across the boundary layer is determined from

$$\mu_t = \min(\mu_{t_i}, \mu_{t_o}) \tag{21-72}$$

that is, μ_t is set to μ_{t_i} from the wall to a location for which μ_{t_i} exceeds the value of μ_{t_o}, at which point μ_t is set to μ_{t_o}.

To model the eddy diffusivity, α_t, or equivalently the turbulent conductivity, the Reynolds analogy may be used. Recall that the Reynolds analogy assumes a similarity between the momentum transfer and heat transfer. Therefore, a turbulent Prandtl number is defined as

$$\text{Pr}_t = \frac{\nu_t}{\alpha_t} = \frac{\mu_t c_p}{k_t} \tag{21-73}$$

For most flows, it is assumed that the turbulent Prandtl number remains constant across the boundary layer. For air, $\text{Pr}_t = 0.9$ for wall bounded flows and 0.5 for shear layers. Thus, the turbulent conductivity is determined as

$$k_t = \frac{\mu_t c_p}{\text{Pr}_t} \tag{21-74}$$

where μ_t is provided by the turbulence models just described.

To initiate the computations, a transition location x_{tr} is specified, and the initial value of the turbulent viscosity is set equal to zero everywhere within the domain. Subsequently, in regions where $x > x_{tr}$, the turbulent viscosity is computed from (21-72) and updated after each time step (or iteration).

21.4.2 One-Equation Turbulence Models

As seen in the previous section, zero-equation models employ an algebraic relation for the eddy viscosity. These models specify length and velocity scales in terms of the mean flow, thus implying an equilibrium between mean motion and turbulence.

One-equation models employ a partial differential equation for velocity scale, whereas the length scale is specified algebraically. The velocity scale is typically written in terms of turbulent kinetic energy k defined as

$$k = \frac{1}{2}(\overline{u'^2} + \overline{v'^2} + \overline{w'^2}) \tag{21-75}$$

The turbulent viscosity μ_t is now written as

$$\mu_t = \rho k^{\frac{1}{2}} \ell \tag{21-76}$$

In the following section, two recently developed one-equation turbulence models are reviewed.

21.4.2.1 Baldwin-Barth One-Equation Turbulence Model:

The Baldwin-Barth one-equation model is obtained from the standard k-ϵ two-equation model. Note that the k-ϵ model has not been introduced as yet, and it will be introduced later on in Section 21.4.3.1. However, since the mathematical details of the Baldwin-Barth one-equation model will not be considered here, knowledge of the k-ϵ model at this point is not necessary. Mathematical details can be found in a report by Baldwin and Barth [21-11].

The procedure to obtain the Baldwin-Barth one-equation model is to combine the k-ϵ two-equation model to provide a single equation in terms of the Turbulence Reynolds number Re_T, where $Re_T = k^2/\nu\epsilon$. The variables in Re_T which are k, ϵ, and ν are, respectively, turbulence kinetic energy given by (21-75), dissipation of turbulence given by (21-152), and kinematic viscosity. To account for the near wall regions, the turbulence Reynolds number is split into two parts as $Re_T = \overline{Re}_T f(\overline{Re}_T)$, where f is a damping function such that $Re_T \cong \overline{Re}_T$ for large Re_T.

The transport equation for turbulence Reynolds number $\overline{Re}_T$ is written as

$$\begin{aligned} \frac{\partial}{\partial t}(\nu \overline{Re}_T) + \vec{V} \cdot \nabla(\nu \overline{Re}_T) &= \\ \left(\nu + \frac{\nu_t}{\sigma_\epsilon}\right) \nabla^2(\nu \overline{Re}_T) - \frac{1}{\sigma_\epsilon}(\nabla \nu_t) \cdot \nabla(\nu \overline{Re}_T) &+ (c_{\epsilon_2} f_2 - c_{\epsilon_1})\sqrt{\nu \overline{Re}_T P} \end{aligned} \tag{21-77}$$

The turbulence production based on Boussinesq assumption is provided from

$$P = \nu_t \left(\frac{\partial u_i}{\partial x_j} + \frac{\partial u_j}{\partial x_i}\right) \frac{\partial u_i}{\partial x_j} - \frac{2}{3}\nu_t \left(\frac{\partial u_k}{\partial x_k}\right)^2 = \nu_t S^2 \tag{21-78}$$

The kinematic eddy viscosity ν_t is determined from

$$\nu_t = c_\mu (\nu \overline{Re}_T) D_1 D_2 \tag{21-79}$$

where D_1 and D_2 are damping functions which extend the validity of the model to near wall regions and are given by

$$D_1 = 1 - \exp(-y^+/A^+) \tag{21-80}$$

$$D_2 = 1 - \exp(-y^+/A_2^+) \tag{21-81}$$

Various functions and constants appearing in Equation (21-77) are as follows.

$$\frac{1}{\sigma_\epsilon} = (c_{\epsilon_2} - c_{\epsilon_1})\sqrt{c_\mu}/\kappa^2$$

$$\mu_t = \rho\nu_t$$

$$f_2(y^+) = \frac{c_{\epsilon_1}}{c_{\epsilon_2}} + \left(1 - \frac{c_{\epsilon_1}}{c_{\epsilon_2}}\right)\left(\frac{1}{\kappa y^+} + D_1 D_2\right)\left[\sqrt{D_1 D_2} + \right.$$

$$\left. \frac{y^+}{\sqrt{D_1 D_2}}\left(\frac{1}{A^+}\exp(-y^+/A^+)D_2 + \frac{1}{A_2^+}\exp(-y^+/A_2^+)D_1\right)\right] \qquad (21\text{-}82)$$

$$\kappa = 0.41, \qquad c_{\epsilon_1} = 1.2, \qquad c_{\epsilon_2} = 2.0$$

$$c_\mu = 0.09, \qquad A^+ = 26, \qquad A_2^+ = 10$$

Equation (21-77) can be written as

$$\frac{\partial}{\partial t}(\nu \overline{Re_T}) + \vec{V}\cdot\nabla(\nu\overline{Re_T}) = \left(\nu + \frac{\nu_t}{\sigma_\epsilon}\right)\nabla^2(\nu\overline{Re_T}) - \frac{1}{\sigma_\epsilon}(\nabla\nu_t)\cdot\nabla(\nu\overline{Re_T})$$

$$+ (c_{\epsilon 2} f_2 - c_{\epsilon 1})\sqrt{c_\mu D_1 D_2} S\nu\overline{Re_T} \qquad (21\text{-}83)$$

where

$$S^2 = 2\left[\left(\frac{\partial u}{\partial x}\right)^2 + \left(\frac{\partial v}{\partial y}\right)^2 + \frac{\partial u}{\partial y}\frac{\partial v}{\partial x}\right] + \left(\frac{\partial u}{\partial y}\right)^2 + \left(\frac{\partial v}{\partial x}\right)^2 - \frac{2}{3}\left(\frac{\partial u}{\partial x} + \frac{\partial v}{\partial y}\right)^2 \qquad (21\text{-}84)$$

as obtained from (21-78) for two-dimensional applications. Due to numerical considerations, Equation (21-83) can be rewritten by manipulation of the diffusion term and with the assumption of constant ν as follows:

$$\frac{\partial}{\partial t}(\nu \overline{Re_T}) + \vec{V}\cdot\nabla(\nu\overline{Re_T}) = 2\left(\nu + \frac{\nu_t}{\sigma_\epsilon}\right)\nabla^2(\nu\overline{Re_T}) - \frac{1}{\sigma_\epsilon}\nabla\cdot\nu_t\nabla(\nu\overline{Re_T})$$

$$+ (c_{\epsilon 2} f_2 - c_{\epsilon 1})\sqrt{c_\mu D_1 D_2} S\nu\overline{Re_T} \qquad (21\text{-}85)$$

Note that the unknown in either Equation (21-83) or (21-85) is $\overline{Re_T}$. Once $\overline{Re_T}$ is determined, relation (21-79) is used to determine ν_t.

21.4.2.1.1 Nondimensional form: Since the governing equations to be solved are typically in nondimensional form, the equations for turbulence models are also nondimensionalized. The same nondimensional variables, as defined in Section 11.3, are used, namely,

$$t^* = \frac{tu_\infty}{L} \quad , \quad x^* = \frac{x}{L} \quad , \quad y^* = \frac{y}{L}$$

$$\mu^* = \frac{\mu}{\mu_\infty} \quad , \quad \rho^* = \frac{\rho}{\rho_\infty} \quad , \quad \nu^* = \frac{\nu}{\nu_\infty}$$

$$\nu_t^* = \frac{\nu_t}{\nu_\infty} \quad , \quad k^* = \frac{k}{u_\infty^2} \quad , \quad \epsilon^* = \epsilon\frac{L}{u_\infty^3}$$

$$Re_T^* = \frac{k^{*^2}}{\nu^*\epsilon^*}$$

Now, the nondimensional form of the Baldwin-Barth one-equation turbulence model corresponding to (21-83) or (21-85) is given by the following equations where the asterisk has been dropped.

$$\begin{aligned}\frac{\partial}{\partial t}(\nu\overline{Re_T}) + \vec{V}\cdot\nabla(\nu\overline{Re_T}) &= \frac{1}{Re_\infty}\left(\nu + \frac{\nu_t}{\sigma_\epsilon}\right)\nabla^2(\nu\overline{Re_T}) - \frac{1}{Re_\infty}\frac{1}{\sigma_\epsilon}(\nabla\nu_t)\cdot\nabla(\nu\overline{Re_T}) \\ &\quad + (c_{\epsilon 2}f_2 - c_{\epsilon 1})\sqrt{c_\mu D_1 D_2 S\nu\overline{Re_T}}\end{aligned} \tag{21-86}$$

or

$$\begin{aligned}\frac{\partial}{\partial t}(\nu\overline{Re_T}) + \vec{V}\cdot\nabla(\nu\overline{Re_T}) &= \frac{2}{Re_\infty}\left(\nu + \frac{\nu_t}{\sigma_\epsilon}\right)\nabla^2(\nu\overline{Re_T}) - \frac{1}{Re_\infty}\frac{1}{\sigma_\epsilon}\nabla\cdot\nu_t\nabla(\nu\overline{Re_T}) \\ &\quad + (c_{\epsilon 2}f_2 - c_{\epsilon 1})\sqrt{c_\mu D_1 D_2}\, S\nu\overline{Re_T}\end{aligned} \tag{21-87}$$

21.4.2.1.2 Baldwin-Barth turbulence model in computational space: In order to numerically solve either Equation (21-86) or (21-87) for a general domain, a coordinate transformation is performed. This transformation is similar to the transformation of the Navier-Stokes equations given in Chapter 11. The expressions for the Cartesian derivatives are

$$\frac{\partial}{\partial x} = \xi_x\frac{\partial}{\partial \xi} + \eta_x\frac{\partial}{\partial \eta} \tag{21-88}$$

$$\frac{\partial}{\partial y} = \xi_y\frac{\partial}{\partial \xi} + \eta_y\frac{\partial}{\partial \eta} \tag{21-89}$$

$$\begin{aligned}\frac{\partial^2}{\partial x^2} &= \xi_x^2\frac{\partial^2}{\partial \xi^2} + 2\xi_x\eta_x\frac{\partial^2}{\partial\xi\partial\eta} + \eta_x^2\frac{\partial^2}{\partial\eta^2} \\ &\quad + \xi_x\left[\frac{\partial\xi_x}{\partial\xi}\frac{\partial}{\partial\xi} + \frac{\partial\eta_x}{\partial\xi}\frac{\partial}{\partial\eta}\right] + \eta_x\left[\frac{\partial\xi_x}{\partial\eta}\frac{\partial}{\partial\xi} + \frac{\partial\eta_x}{\partial\eta}\frac{\partial}{\partial\eta}\right]\end{aligned} \tag{21-90}$$

$$\frac{\partial^2}{\partial y^2} = \xi_y^2 \frac{\partial^2}{\partial \xi^2} + 2\xi_y \eta_y \frac{\partial^2}{\partial \xi \partial \eta} + \eta_y^2 \frac{\partial^2}{\partial \eta^2}$$

$$+ \xi_y \left[\frac{\partial \xi_y}{\partial \xi} \frac{\partial}{\partial \xi} + \frac{\partial \eta_y}{\partial \xi} \frac{\partial}{\partial \eta} \right] + \eta_y \left[\frac{\partial \xi_y}{\partial \eta} \frac{\partial}{\partial \xi} + \frac{\partial \eta_y}{\partial \eta} \frac{\partial}{\partial \eta} \right] \tag{21-91}$$

Now, for simplicity, define

$$\nu \overline{Re_T} = R$$

Then, the transformed Baldwin-Barth turbulence model given by Equation (21-86) is written as follows. The details of transformation and mathematical manipulation is provided in Appendix I.

$$\frac{\partial R}{\partial t} = -U \frac{\partial R}{\partial \xi} - V \frac{\partial R}{\partial \eta} + \frac{1}{Re_\infty} \left(\nu + \frac{\nu_t}{\sigma_\epsilon} \right) \left[a_4 \frac{\partial^2 R}{\partial \xi^2} + 2c_5 \frac{\partial^2 R}{\partial \xi \partial \eta} + b_4 \frac{\partial^2 R}{\partial \eta^2} \right.$$

$$\left. + g_1 \frac{\partial R}{\partial \xi} + g_2 \frac{\partial R}{\partial \eta} \right] - \frac{1}{Re_\infty} \frac{1}{\sigma_\epsilon} \left[a_4 \frac{\partial \nu_t}{\partial \xi} \frac{\partial R}{\partial \xi} + c_5 \left(\frac{\partial \nu_t}{\partial \xi} \frac{\partial R}{\partial \eta} + \frac{\partial \nu_t}{\partial \eta} \frac{\partial R}{\partial \xi} \right) + b_4 \frac{\partial \nu_t}{\partial \eta} \frac{\partial R}{\partial \eta} \right]$$

$$+ (c_{\epsilon 2} f_2 - c_{\epsilon 1}) \sqrt{c_\mu D_1 D_2}\, S R \tag{21-92}$$

where U and V are the contravarient velocity components

$$U = \xi_x u + \xi_y v$$

$$V = \eta_x u + \eta_y v$$

and

$$a_4 = \xi_x^2 + \xi_y^2 \tag{21-93}$$

$$b_4 = \eta_x^2 + \eta_y^2 \tag{21-94}$$

$$c_5 = \xi_x \eta_x + \xi_y \eta_y \tag{21-95}$$

$$g_1 = \xi_x \frac{\partial \xi_x}{\partial \xi} + \eta_x \frac{\partial \xi_x}{\partial \eta} + \xi_y \frac{\partial \xi_y}{\partial \xi} + \eta_y \frac{\partial \xi_y}{\partial \eta} \tag{21-96}$$

$$g_2 = \xi_x \frac{\partial \eta_x}{\partial \xi} + \eta_x \frac{\partial \eta_x}{\partial \eta} + \xi_y \frac{\partial \eta_y}{\partial \xi} + \eta_y \frac{\partial \eta_y}{\partial \eta} \tag{21-97}$$

Furthermore, the expression for S in the last term of Equation (21-92) is

$$S^2 = a_1 \left(\frac{\partial u}{\partial \xi} \right)^2 + b_1 \left(\frac{\partial u}{\partial \eta} \right)^2 + a_2 \left(\frac{\partial v}{\partial \xi} \right)^2 + b_2 \left(\frac{\partial v}{\partial \eta} \right)^2$$

$$+ 2c_1 \left(\frac{\partial u}{\partial \xi} \frac{\partial u}{\partial \eta} \right) + 2c_2 \left(\frac{\partial v}{\partial \xi} \frac{\partial v}{\partial \eta} \right) + 2a_3 \frac{\partial u}{\partial \xi} \frac{\partial v}{\partial \xi}$$

$$+ 2b_3 \frac{\partial u}{\partial \eta} \frac{\partial v}{\partial \eta} + 2c_3 \frac{\partial u}{\partial \xi} \frac{\partial v}{\partial \eta} + 2c_4 \frac{\partial u}{\partial \eta} \frac{\partial v}{\partial \xi} \tag{21-98}$$

where the coefficients a_1, a_2, a_3, b_1, b_2, b_3, c_1, c_2, c_3 and c_4 are given by (11-210a), (11-210b), (11-210c), (11-211a), (11-211b), (11-211c), (11-212a), (11-212b), (11-212c), and (11-212d), respectively. The transformation of Equation (21-87) is carried out in a slightly different fashion, the details of which are given in Appendix I. The transformed equation is given as

$$\begin{aligned} \frac{\partial R}{\partial t} &= -U \frac{\partial R}{\partial \xi} - V \frac{\partial R}{\partial \eta} + \frac{2}{Re_\infty} \left(\nu + \frac{\nu_t}{\sigma_\epsilon} \right) \left[\xi_x \frac{\partial A}{\partial \xi} + \xi_y \frac{\partial B}{\partial \xi} + \eta_x \frac{\partial A}{\partial \eta} + \eta_y \frac{\partial B}{\partial \eta} \right] \\ &\quad - \frac{1}{Re_\infty} \frac{1}{\sigma_\epsilon} \left[\xi_x \frac{\partial \alpha}{\partial \xi} + \xi_y \frac{\partial \beta}{\partial \xi} + \eta_x \frac{\partial \alpha}{\partial \eta} + \eta_y \frac{\partial \beta}{\partial \eta} \right] \\ &\quad + (c_{\epsilon 2} f_2 - c_{\epsilon 1}) \sqrt{c_\mu D_1 D_2}\, SR \end{aligned} \tag{21-99}$$

where

$$A = \frac{\partial R}{\partial x} = \xi_x \frac{\partial R}{\partial \xi} + \eta_x \frac{\partial R}{\partial \eta} \tag{21-100}$$

$$B = \frac{\partial R}{\partial y} = \xi_y \frac{\partial R}{\partial \xi} + \eta_y \frac{\partial R}{\partial \eta} \tag{21-101}$$

$$\alpha = \nu_t \left(\xi_x \frac{\partial R}{\partial \xi} + \eta_x \frac{\partial R}{\partial \eta} \right) \tag{21-102}$$

$$\beta = \nu_t \left(\xi_y \frac{\partial R}{\partial \xi} + \eta_y \frac{\partial R}{\partial \eta} \right) \tag{21-103}$$

In order to simplify the development of finite difference equations, Equation (21-99) is expressed as

$$\frac{\partial R}{\partial t} = M + P \tag{21-104}$$

The term M, which represents the convection and diffusion of turbulence can be written as

$$M = M^1 + M_\xi^2 + M_\eta^2 + M_\xi^3 + M_\eta^3 \tag{21-105}$$

where

$$M^1 = -U \frac{\partial R}{\partial \xi} - V \frac{\partial R}{\partial \eta} \tag{21-106}$$

$$M_\xi^2 = -\frac{1}{Re_\infty}\frac{1}{\sigma_\epsilon}\left(\xi_x\frac{\partial\alpha}{\partial\xi} + \xi_y\frac{\partial\beta}{\partial\xi}\right) \tag{21-107}$$

$$M_\eta^2 = -\frac{1}{Re_\infty}\frac{1}{\sigma_\epsilon}\left(\eta_x\frac{\partial\alpha}{\partial\eta} + \eta_y\frac{\partial\beta}{\partial\eta}\right) \tag{21-108}$$

$$M_\xi^3 = \frac{2}{Re_\infty}\left(\nu + \frac{\nu_t}{\sigma_\epsilon}\right)\left(\xi_x\frac{\partial A}{\partial\xi} + \xi_y\frac{\partial B}{\partial\xi}\right) \tag{21-109}$$

$$M_\eta^3 = \frac{2}{Re_\infty}\left(\nu + \frac{\nu_t}{\sigma_\epsilon}\right)\left(\eta_x\frac{\partial A}{\partial\eta} + \eta_y\frac{\partial B}{\partial\eta}\right) \tag{21-110}$$

and

$$P = (c_{\epsilon 2}f_2 - c_{\epsilon 1})\sqrt{c_\mu D_1 D_2}\, SR$$

The initial and boundary conditions are set according to:

(1) Initial condition: A value of $\overline{Re_T} = (\overline{Re_T})_\infty < 1$ is specified within the domain to initiate the solution.

(2) Boundary conditions: The value of $\overline{Re_T}$ is set equal to zero at the solid surface, and it is set to its initial value of $(\overline{Re_T})_\infty$ at the inflow. At the outflow, extrapolation from the interior domain is used to specify $\overline{Re_T}$ at the boundary.

21.4.2.2 Spalart-Allmaras One-Equation Turbulence Model: The Spalart-Allmaras model solves a transport equation for a working variable $\overline{\nu}$ which is related to the eddy viscosity. The governing equation is derived by using empiricism, dimensional analysis, Galilean invariance, and selected dependence on the molecular viscosity [21-12]. The transport equation is expressed as

$$\frac{d\overline{\nu}}{dt} = \frac{1}{\sigma}\left[\nabla\cdot\left((\nu+\overline{\nu})\nabla\overline{\nu}\right) + c_{b2}(\nabla\overline{\nu})^2\right] + c_{b1}\bar{S}\overline{\nu}(1 - f_{t2})$$

$$- \left[c_{w1}f_w - \frac{c_{b1}}{\kappa^2}f_{t2}\right]\left[\frac{\overline{\nu}}{d}\right]^2 + f_{t1}(\Delta q)^2 \tag{21-111}$$

where the eddy viscosity is given by

$$\nu_t = \overline{\nu} f_{v1} \tag{21-112}$$

and

$$f_{v1} = \frac{\chi^3}{\chi^3 + c_{v1}^3} \tag{21-113}$$

$$\chi = \frac{\overline{\nu}}{\nu} \tag{21-114}$$

Various functions and constants appearing in Equation (21-111) are defined as

$$\bar{S} = S + \frac{\overline{\nu}}{\kappa^2 d^2} f_{v2} \tag{21-115}$$

where d is the distance to the wall, κ is the von Karman constant, S is the magnitude of the vorticity,

$$S = \left| \frac{\partial v}{\partial x} - \frac{\partial u}{\partial y} \right| \tag{21-116}$$

and

$$f_{v2} = 1 - \frac{\chi}{1 + \chi f_{v1}} \tag{21-117}$$

It should be noted that expression (21-66) has also been used for S. The function f_w is

$$f_w(r) = g \left(\frac{1 + c_{w3}^6}{g^6 + c_{w3}^6} \right)^{\frac{1}{6}} \tag{21-118}$$

where

$$g = r + c_{w2}(r^6 - r) \tag{21-119}$$

and

$$r = \frac{\overline{\nu}}{\bar{S} \kappa^2 d^2} \tag{21-120}$$

Large values of r should be truncated to a value of about 10. The function f_{t2} is given by

$$f_{t2} = c_{t3} \exp(-c_{t4} \chi^2) \tag{21-121}$$

and the trip function f_{t1} is

$$f_{t1} = c_{t1} g_t \exp \left[-c_{t2} \left(\frac{\omega_t}{\Delta q} \right)^2 (d^2 + g_t^2 d_t^2) \right] \tag{21-122}$$

The following are used in Equation (21-122):
d_t: The distance from the field point to the trip which is located on the surface.
ω_t: The wall vorticity at the trip.
Δq: The difference between the velocities at the field point and trip.
g_t: $g_t = \min[1.0, \Delta q / \omega_t \Delta x]$, where Δx is the grid spacing along the wall at the trip.

The constants used in the equations above are

$\sigma = \frac{2}{3}$ $\quad c_{b1} = 0.1355$ $\quad c_{b2} = 0.622$

$c_{w1} = \frac{c_{b1}}{\kappa^2} + (1 + c_{b2})/\sigma$ $\quad c_{w2} = 0.3$ $\quad c_{w3} = 2$ $\quad \kappa = 0.41$

$c_{v1} = 7.1$ $\quad c_{t1} = 1.0$ $\quad c_{t2} = 2.0$ $\quad c_{t3} = 1.1$ $\quad c_{t4} = 2.0$

The Spalart-Allmaras turbulence model given by (21-111) can also be written as

$$\frac{d\overline{\nu}}{dt} = \left(\frac{1+c_{b2}}{\sigma}\right)\nabla\cdot[(\nu+\overline{\nu})\nabla\overline{\nu}] - \frac{c_{b2}}{\sigma}(\nu+\overline{\nu})\nabla^2\overline{\nu} + c_{b1}(1-f_{t2})\overline{\nu}\overline{S} - \left[c_{w1}f_w - \frac{c_{b1}}{\kappa^2}f_{t2}\right]\left[\frac{\overline{\nu}}{d}\right]^2 + f_{t1}(\Delta q)^2 \tag{21-123}$$

The first two terms on the right-hand side of Equation (21-123) can be expanded and recombined to provide the following equation.

$$\frac{d\overline{\nu}}{dt} = \left(\frac{1+c_{b2}}{\sigma}\right)[\nabla(\nu+\overline{\nu})\cdot\nabla\overline{\nu}] + \frac{1}{\sigma}(\nu+\overline{\nu})\nabla^2\overline{\nu} + c_{b1}(1-f_{t2})\overline{\nu}\overline{S} - \left[c_{w1}f_w - \frac{c_{b1}}{\kappa^2}f_{t2}\right]\left[\frac{\overline{\nu}}{d}\right]^2 + f_{t1}(\Delta q)^2 \tag{21-124}$$

Either one of Equations (21-123) or (21-124) can be used.

The initial condition for $\bar{\nu}$ is specified to be zero up to a value of $\nu_\infty/10$, that is, $\bar{\nu} = 0 \rightarrow \nu_\infty/10$. The boundary conditions are set according to:

(1) At the inflow, $\bar{\nu} = \bar{\nu}_\infty$,

(2) At the solid surface $\bar{\nu} = 0$, and

(3) At the outflow, extrapolation is used.

21.4.2.2.1 Nondimensional form: Using the same nondimensional variables defined previously, the nondimensionalized Spalart-Allmaras turbulence model is given by the following equation. Again, as in the previous model, the notation of asterisk has been dropped.

$$\frac{d\overline{\nu}}{dt} = \left(\frac{1}{Re_\infty}\right)\left(\frac{1+c_{b2}}{\sigma}\right)\nabla\cdot[(\nu+\overline{\nu})\nabla\overline{\nu}] - \left(\frac{1}{Re_\infty}\right)\left(\frac{c_{b2}}{\sigma}\right)(\nu+\overline{\nu})\nabla^2\overline{\nu} + c_{b1}(1-f_{t2})S\overline{\nu} - \left(\frac{1}{Re_\infty}\right)(c_{w1}f_w)\left(\frac{\overline{\nu}}{d}\right)^2 + \left(\frac{1}{Re_\infty}\right)\left(\frac{c_{b1}}{\kappa^2}\right)[(1-f_{t2})f_{v2}+f_{t2}]\left(\frac{\overline{\nu}}{d}\right)^2 + Re_\infty f_{t1}(\Delta q)^2 \tag{21-125}$$

which corresponds to Equation (21-123) and

$$\frac{d\overline{\nu}}{dt} = \left(\frac{1}{Re_\infty}\right)\left(\frac{1+c_{b2}}{\sigma}\right)[\nabla(\nu+\overline{\nu})\cdot\nabla\overline{\nu}] + \left(\frac{1}{Re_\infty}\right)\frac{1}{\sigma}(\nu+\overline{\nu})\nabla^2\overline{\nu} + c_{b1}(1-f_{t2})\overline{\nu}\overline{S} - \left(\frac{1}{Re_\infty}\right)(c_{w1}f_w)\left(\frac{\overline{\nu}}{d}\right)^2$$

$$+\left(\frac{1}{Re_\infty}\right)\left(\frac{c_{b1}}{\kappa^2}\right)[(1-f_{t2})f_{v2}+f_{t2}]\left(\frac{\overline{\nu}}{d}\right)^2+Re_\infty f_{t1}(\Delta q)^2 \qquad (21\text{-}126)$$

corresponding to Equation (21-124).

21.4.2.2.2 Spalart-Allmaras turbulence model in computational space: Using the expressions for Cartesian derivatives given by (21-80) through (21-91), the Spalart-Allmaras turbulence model given by (21-126) in generalized coordinates is written as

$$\frac{\partial\overline{\nu}}{\partial t}+U\frac{\partial\overline{\nu}}{\partial\xi}+V\frac{\partial\overline{\nu}}{\partial\eta}=\left(\frac{1}{Re_\infty}\right)\left(\frac{1+c_{b2}}{\sigma}\right)\left\{a_4\frac{\partial\overline{\nu}}{\partial\xi}\frac{\partial}{\partial\xi}(\nu+\overline{\nu})+b_4\frac{\partial\overline{\nu}}{\partial\eta}\frac{\partial}{\partial\eta}(\nu+\overline{\nu})\right.$$

$$\left.+c_5\left[\frac{\partial\overline{\nu}}{\partial\eta}\frac{\partial}{\partial\xi}(\nu+\overline{\nu})+\frac{\partial\overline{\nu}}{\partial\xi}\frac{\partial}{\partial\eta}(\nu+\overline{\nu})\right]\right\}+\left(\frac{1}{Re_\infty}\right)\left(\frac{1}{\sigma}\right)(\nu+\overline{\nu})\left[a_4\frac{\partial^2\overline{\nu}}{\partial\xi^2}\right.$$

$$\left.+2c_5\frac{\partial^2\overline{\nu}}{\partial\xi\partial\eta}+b_4\frac{\partial^2\overline{\nu}}{\partial\eta^2}+g_1\frac{\partial\overline{\nu}}{\partial\xi}+g_2\frac{\partial\overline{\nu}}{\partial\eta}\right]+c_{b1}(1-f_{t2})S\overline{\nu}-\left(\frac{1}{Re_\infty}\right)(c_{w1}f_w)\left(\frac{\overline{\nu}}{d}\right)^2$$

$$+\left(\frac{1}{Re_\infty}\right)\left(\frac{c_{b1}}{\kappa^2}\right)\left[(1-f_{t2})f_{v2}+f_{t2}\right]\left(\frac{\overline{\nu}}{d}\right)^2+Re_\infty f_{t1}(\Delta q)^2 \qquad (21\text{-}127)$$

where

$$S=\left|\xi_y\frac{\partial u}{\partial\xi}+\eta_y\frac{\partial u}{\partial\eta}-\xi_x\frac{\partial v}{\partial\xi}-\eta_x\frac{\partial v}{\partial\eta}\right| \qquad (21\text{-}128)$$

and, as defined previously,

$$a_4 = \xi_x^2+\xi_y^2 \qquad (21\text{-}129)$$

$$b_4 = \eta_x^2+\eta_y^2 \qquad (21\text{-}130)$$

$$c_5 = \xi_x\eta_x+\xi_y\eta_y \qquad (21\text{-}131)$$

$$g_1 = \xi_x\frac{\partial\xi_x}{\partial\xi}+\eta_x\frac{\partial\xi_x}{\partial\eta}+\xi_y\frac{\partial\xi_y}{\partial\xi}+\eta_y\frac{\partial\xi_y}{\partial\eta} \qquad (21\text{-}132)$$

$$g_2 = \xi_x\frac{\partial\eta_x}{\partial\xi}+\eta_x\frac{\partial\eta_x}{\partial\eta}+\xi_y\frac{\partial\eta_y}{\partial\xi}+\eta_y\frac{\partial\eta_y}{\partial\eta} \qquad (21\text{-}133)$$

The transformed Spalart-Allmaras turbulence model given by (21-125) is

$$\frac{\partial\overline{\nu}}{\partial t}+U\frac{\partial\overline{\nu}}{\partial\xi}+V\frac{\partial\overline{\nu}}{\partial\eta}=\left(\frac{1}{Re_\infty}\right)\left(\frac{1+c_{b2}}{\sigma}\right)\left\{\xi_x\frac{\partial\alpha}{\partial\xi}+\eta_x\frac{\partial\alpha}{\partial\eta}+\xi_y\frac{\partial\beta}{\partial\xi}+\eta_y\frac{\partial\beta}{\partial\eta}\right\}$$

$$- \left(\frac{1}{Re_\infty}\right) \frac{c_{b2}}{\sigma}(\nu+\overline{\nu}) \left\{\xi_x\frac{\partial A}{\partial \xi} + \eta_x\frac{\partial A}{\partial \eta} + \xi_y\frac{\partial B}{\partial \xi} + \eta_y\frac{\partial B}{\partial \eta}\right\}$$

$$+ \left(\frac{1}{Re_\infty}\right) \left(\frac{c_{b1}}{\kappa^2}\right) [(1-f_{t2})f_{v2} + f_{t2}] \left(\frac{\overline{\nu}}{d}\right)^2 - \left(\frac{1}{Re_\infty}\right) (c_{w1} f_w) \left(\frac{\overline{\nu}}{d}\right)^2$$

$$+ c_{b1}(1-f_{t2})S\overline{\nu} + Re_\infty f_{t1}(\Delta q)^2 \tag{21-134}$$

where

$$A = \frac{\partial \overline{\nu}}{\partial x} = \xi_x\frac{\partial \overline{\nu}}{\partial \xi} + \eta_x\frac{\partial \overline{\nu}}{\partial \eta} \tag{21-135}$$

$$B = \frac{\partial \overline{\nu}}{\partial y} = \xi_y\frac{\partial \overline{\nu}}{\partial \xi} + \eta_y\frac{\partial \overline{\nu}}{\partial \eta} \tag{21-136}$$

$$\alpha = (\nu+\overline{\nu}) \left(\xi_x\frac{\partial \overline{\nu}}{\partial \xi} + \eta_x\frac{\partial \overline{\nu}}{\partial \eta}\right) \tag{21-137}$$

$$\beta = (\nu+\overline{\nu}) \left(\xi_y\frac{\partial \overline{\nu}}{\partial \xi} + \eta_y\frac{\partial \overline{\nu}}{\partial \eta}\right) \tag{21-138}$$

Equation (21-134) is now written as

$$\frac{\partial \overline{\nu}}{\partial t} = M + P + D + T \tag{21-139}$$

where

(a) The term M represents the convection and diffusion of turbulence and, due to numerical consideration, is decomposed as follows:

$$M = M^1 + M^2 + M^3 \tag{21-140}$$

Each term in expression (21-140) is defined as

$$M^1 = -\left(U\frac{\partial \overline{\nu}}{\partial \xi} + V\frac{\partial \overline{\nu}}{\partial \eta}\right) \tag{21-141}$$

$$M^2 = \left(\frac{1}{Re_\infty}\right) \left(\frac{1+c_{b2}}{\sigma}\right) \left\{\xi_x\frac{\partial \alpha}{\partial \xi} + \eta_x\frac{\partial \alpha}{\partial \eta} + \xi_y\frac{\partial \beta}{\partial \xi} + \eta_y\frac{\partial \beta}{\partial \eta}\right\} \tag{21-142}$$

$$M^3 = -\left(\frac{1}{Re_\infty}\right) \frac{c_{b2}}{\sigma}(\nu+\overline{\nu}) \left\{\xi_x\frac{\partial A}{\partial \xi} + \eta_x\frac{\partial A}{\partial \eta} + \xi_y\frac{\partial B}{\partial \xi} + \eta_y\frac{\partial B}{\partial \eta}\right\} \tag{21-143}$$

The terms M^2 and M^3 can be further regrouped as terms including ξ derivatives and terms involving η derivatives as follow

$$M_\xi^2 = \left(\frac{1}{Re_\infty}\right)\left(\frac{1+c_{b2}}{\sigma}\right)\left(\xi_x\frac{\partial\alpha}{\partial\xi}+\xi_y\frac{\partial\beta}{\partial\xi}\right) \tag{21-144}$$

$$M_\eta^2 = \left(\frac{1}{Re_\infty}\right)\left(\frac{1+c_{b2}}{\sigma}\right)\left(\eta_x\frac{\partial\alpha}{\partial\eta}+\eta_y\frac{\partial\beta}{\partial\eta}\right) \tag{21-145}$$

$$M_\xi^3 = -\left(\frac{1}{Re_\infty}\right)\frac{c_{b2}}{\sigma}(\nu+\overline{\nu})\left(\xi_x\frac{\partial A}{\partial\xi}+\xi_y\frac{\partial B}{\partial\xi}\right) \tag{21-146}$$

$$M_\eta^3 = -\left(\frac{1}{Re_\infty}\right)\frac{c_{b2}}{\sigma}(\nu+\overline{\nu})\left(\eta_x\frac{\partial A}{\partial\eta}+\eta_y\frac{\partial B}{\partial\eta}\right) \tag{21-147}$$

(b) The second term in Equation (21-139) represents the production of turbulence and is

$$\begin{aligned} P &= c_{b1}(1-f_{t2})S\overline{\nu}+\left(\frac{1}{Re_\infty}\right)\frac{c_{b1}}{\kappa^2}(1-f_{t2})f_{v2}\left[\frac{\overline{\nu}}{d}\right]^2 \\ &= c_{b1}(1-f_{t2})\overline{\nu}\left[S+\frac{\overline{\nu}}{Re_\infty\kappa^2 d^2}f_{v2}\right] = c_{b1}(1-f_{t2})\overline{\nu}\,\overline{S} \end{aligned} \tag{21-148}$$

Note that $\overline{S} = S + \dfrac{1}{Re_\infty}\left(\dfrac{\overline{\nu}}{\kappa^2 d^2}\right) f_{v2}$

(c) The third term D represents the destruction of turbulence and is given by

$$D = -\left(\frac{1}{Re_\infty}\right)\left(c_{w1}f_w-\frac{c_{b1}}{\kappa^2}f_{t2}\right)\left[\frac{\overline{\nu}}{d}\right]^2 \tag{21-149}$$

(d) And, finally, the last term in Equation (21-139) represents the trip term which causes/ triggers transition from laminar flow to turbulent flow.

$$T = (Re_\infty)f_{t1}(\Delta q)^2 \tag{21-150}$$

21.4.3 Two-Equation Turbulence Models

It was pointed out previously that the convection of turbulence is not modeled in zero-equation models. Therefore, the physical effect of past history of the flow is not included in simple algebraic models. In order to account for this physical effect, a transport equation based on the Navier-Stokes equation may be derived. When one such equation is employed, it is referred to as a one-equation model which was discussed in the previous section. When two transport equations are used, it is known as a *two-equation model.*

Complex flowfields which include massively separated flows, unsteadiness, and flows involving multiple-length scales occur frequently in fluid mechanics applications. In these types of complex flows, the lower order turbulence models, that is, zero-, half-, or one-equation models, become very complicated and often ambiguous. Two-equation models are developed to better represent the physics of turbulence in these types of complex flowfields. In this section, k-ϵ and k-ω two-equation turbulence models are reviewed.

21.4.3.1 k-ε Two-Equation Turbulence Model:

A commonly used two-equation turbulence model is the k-ϵ model. The partial differential equations are derived for kinetic energy of turbulence (k), and the dissipation of turbulence (ϵ), where

$$k = \frac{1}{2}\left[\overline{u'^2} + \overline{v'^2} + \overline{w'^2}\right] \tag{21-151}$$

and

$$\epsilon = \nu_t \overline{\left(\frac{\partial u_i'}{\partial x_j}\right)\left(\frac{\partial u_i'}{\partial x_j}\right)} \tag{21-152}$$

Since it is important to identify the terms involved in these equations, and, for the benefit of those who are not familiar with the origin and derivation of these equations, the details of the kinetic energy of turbulence equation and interpretation of terms are provided in Appendix K. Now, the standard k-ϵ two-equation model is expressed by the turbulent kinetic equation

$$\rho\frac{dk}{dt} = \frac{\partial}{\partial x_j}\left[\left(\mu + \frac{\mu_t}{\sigma_k}\right)\frac{\partial k}{\partial x_j}\right] + P_k - \rho\epsilon \tag{21-153}$$

and the dissipation rate equation

$$\rho\frac{d\epsilon}{dt} = \frac{\partial}{\partial x_j}\left[\left(\mu + \frac{\mu_t}{\sigma_\epsilon}\right)\frac{\partial \epsilon}{\partial x_j}\right] + c_{\epsilon 1}P_k\frac{\epsilon}{k} - c_{\epsilon 2}\rho\frac{\epsilon^2}{k} \tag{21-154}$$

where P_k is the production of turbulence defined as $P_k = \tau_{ij}(\partial u_i/\partial x_j)$.

The turbulent kinetic equation and the dissipation rate equation given by (21-153) and (21-154) can be written in an expanded form in Cartesian coordinates for two-dimensional problems as

$$\rho\left(\frac{\partial k}{\partial t} + u\frac{\partial k}{\partial x} + v\frac{\partial k}{\partial y}\right) = \frac{\partial}{\partial x}\left[\left(\mu + \frac{\mu_t}{\sigma_k}\right)\frac{\partial k}{\partial x}\right] + \frac{\partial}{\partial y}\left[\left(\mu + \frac{\mu_t}{\sigma_k}\right)\frac{\partial k}{\partial y}\right] + P_k - \rho\epsilon \tag{21-155}$$

and

$$\rho\left(\frac{\partial \epsilon}{\partial t}+u\frac{\partial \epsilon}{\partial x}+v\frac{\partial \epsilon}{\partial y}\right) = \frac{\partial}{\partial x}\left[\left(\mu+\frac{\mu_t}{\sigma_\epsilon}\right)\frac{\partial \epsilon}{\partial x}\right]+\frac{\partial}{\partial y}\left[\left(\mu+\frac{\mu_t}{\sigma_\epsilon}\right)\frac{\partial \epsilon}{\partial y}\right] + c_{\epsilon 1}P_k\frac{\epsilon}{k} - c_{\epsilon 2}\rho\frac{\epsilon^2}{k} \tag{21-156}$$

where typical constants are

$$\sigma_k = 1.0 \quad \sigma_\epsilon = 1.3 \quad c_{\epsilon 1} = 1.44 \quad c_{\epsilon 2} = 1.92$$

and the turbulent viscosity is related to ϵ by

$$\mu_t = \rho c_\mu \frac{k^2}{\epsilon} \tag{21-157}$$

where $c_\mu = 0.09$.

Before proceeding further, it is important to note that the left-hand side of the k and ϵ equations, given by (21-155) and (21-156), can also be written as

$$\frac{\partial}{\partial t}(\bar{\rho}\,k)+\frac{\partial}{\partial x}(\bar{\rho}\,\bar{u}\,k)+\frac{\partial}{\partial y}(\bar{\rho}\,\bar{v}\,k) = \mathrm{RHS}_k \tag{21-158}$$

and

$$\frac{\partial}{\partial t}(\bar{\rho}\,\epsilon)+\frac{\partial}{\partial x}(\bar{\rho}\,\bar{u}\,\epsilon)+\frac{\partial}{\partial y}(\bar{\rho}\,\bar{v}\,\epsilon) = \mathrm{RHS}_\epsilon \tag{21-159}$$

which is simply obtained by multiplying the continuity equation by k, adding it to the left-hand side of (21-155), and subsequently combining the terms. A similar approach is used for the ϵ-equation.

As for the zero-equation models, a two-layer approach is incorporated. That is, Equations (21-155) and (21-156) are used in the outer region down to the vicinity of the viscous sublayer. For a typical wall flow, that corresponds to a y^+ range of 30–50. Adjacent to the surface, wall functions such as (21-56) are employed. The values of k and ϵ must be provided at the interface as the inner boundary conditions for Equations (21-155) and (21-156). These values may be obtained by the following: if y_i denotes the interface between the outer region and the viscous sublayer, then

$$k_{y=y_i} = \frac{\tau_i}{\rho c_D^{\frac{3}{2}}} \tag{21-160}$$

where c_D is typically 0.164, and τ_i is the shear stress at y_i location. Similarly,

$$\epsilon_{y=y_i} = \frac{c_D k_{y=y_i}^{\frac{3}{2}}}{l} \tag{21-161}$$

where the length scale may be evaluated by the simple linear relation for the viscous sublayer given by

$$l = \kappa y \tag{21-162}$$

Once the local values of the kinetic energy of turbulence and dissipation of turbulence have been computed, turbulent viscosity given by (21-157) may be determined. It is noted that correlations such as $\overline{u'^2}$, $\overline{v'^2}$, and $\overline{w'^2}$ can be determined by semi-empirical relations. For example,

$$\overline{u'^2} = 2a_2 k \tag{21-163}$$

$$\overline{v'^2} = 2a_3 k \tag{21-164}$$

$$\overline{w'^2} = 2(1 - a_2 - a_3)k \tag{21-165}$$

where a_2 and a_3 are structural scales usually assigned values of 0.556 and 0.15, respectively.

21.4.3.1.1 Low Reynolds number k-ε model: The difficulty with the standard k-ϵ model introduced in the previous section is that the equations become numerically unstable when integrated to the wall. To overcome this problem, a two-layer approach is implemented, as discussed previously. However, a better approach is perhaps to directly integrate the k-ϵ equations through the viscous sublayer all the way to the wall. In order to enable the integration to the wall and to improve the capability of the standard k-ϵ model, several modifications are introduced. The resulting formulation is known as the *low-Reynolds number k-ε equations.* The first low-Reynolds number k-ϵ model was developed by Jones and Launder [21-13, 21-14], and subsequently it has been modified by several investigators. The primary modifications introduced by Jones and Launder were to include turbulence Reynolds number dependency functions f_1 and f_2 in Equation (21-154) and f_μ in relation (21-157). Furthermore, additional terms L_k and L_ϵ were added to the equations to account for the dissipation processes which may not be isotropic. Thus the low-Reynolds number k-ϵ equation is written as

$$\rho \frac{dk}{dt} = \frac{\partial}{\partial x_j}\left[\left(\mu + \frac{\mu_t}{\sigma_k}\right)\frac{\partial k}{\partial x_j}\right] + P_k - \rho\epsilon + L_k \tag{21-166}$$

and

$$\rho \frac{d\epsilon}{dt} = \frac{\partial}{\partial x_j}\left[\left(\mu + \frac{\mu_t}{\sigma_\epsilon}\right)\frac{\partial \epsilon}{\partial x_j}\right] + c_{\epsilon 1} f_1 P_k \frac{\epsilon}{k} - c_{\epsilon 2} f_2 \rho \frac{\epsilon^2}{k} + L_\epsilon \tag{21-167}$$

where the turbulent viscosity is now computed according to

$$\mu_t = \rho f_\mu c_\mu \frac{k^2}{\epsilon} \tag{21-168}$$

The following relations are proposed by Jones and Launder.

$$f_1 = 1.0$$

$$f_2 = 1 - 0.3\exp(-Re_T^2)$$

$$f_\mu = \exp\left[\frac{-2.5}{1 + 0.02Re_T}\right]$$

$$L_k = -2\mu\left(\frac{\partial\sqrt{k}}{\partial x_j}\right)^2$$

and

$$L_\epsilon = 2\frac{\mu\,\mu_t}{\rho}\left(\frac{\partial^2 u_i}{\partial x_j^2}\right)^2$$

Since the early 1970's several modifications to the low Reynolds number terms and the model constants have been introduced. A summary of selected k-ϵ models are provided in Table 21.1.

Model	Ref.No.	f_1	f_2	f_μ
Standard		1.0	1.0	1.0
Jones-Launder	12-13 12-14	1.0	$1 - 0.3\exp(-Re_T^2)$	$\exp\left[\frac{-2.5}{1+0.02Re_T}\right]$
Hoffman	12-15	1.0	$1 - 0.3\exp(-Re_T^2)$	$\exp\left[\frac{-1.75}{1+0.02Re_T}\right]$
Nagano-Hishida	12-16	1.0	$1 - 0.3\exp(-Re_T^2)$	$[1 - \exp(-Re_T/26.5)]^2$

Model	L_k	L_ϵ	C_μ	$C_{\epsilon 1}$	$C_{\epsilon 2}$	σ_k	σ_ϵ
Standard	0	0	0.09	1.44	1.92	1.0	1.30
Jones-Launder	$-2\mu\left(\frac{\partial\sqrt{k}}{\partial x_j}\right)^2$	$2\frac{\mu\,\mu_t}{\rho}\left(\frac{\partial^2 u_i}{\partial x_j}\right)^2$	0.09	1.44	1.92	1.0	1.30
Hoffman	$-\frac{\nu}{y}\frac{\partial k}{\partial y}$	0	0.09	1.81	2.0	2.0	3.0
Nagano-Hishida	$-2\nu\left(\frac{\partial\sqrt{k}}{\partial y}\right)^2$	$\nu\,\nu_t(1-f_\mu)\left(\frac{\partial^2 u}{\partial y^2}\right)^2$	0.09	1.45	1.90	1.0	1.30

Table 21.1 Functions and constants for the k-ϵ turbulence models.

21.4.3.1.2 Compressibility correction and axisymmetric consideration: In addition to the modifications of the low Reynolds number terms and model constants shown in Table 21.1, other modifications have also been introduced into the k-ϵ equations, most notably, the compressibiility correction terms and axisymmetric considerations.

The compressibility correction term is typically included by introduction of the turbulent Mach number M_t defined by

$$M_t = \sqrt{\frac{2k}{a^2}}$$

Now, the k-ϵ equations including the compressibility correction terms are written as

$$\frac{\partial}{\partial t}(\bar{\rho}k) + \frac{\partial}{\partial x_j}(\bar{\rho}\bar{u}_j k) = \frac{\partial}{\partial x_j}\left[\left(\mu + \frac{\mu_t}{\sigma_k}\right)\frac{\partial k}{\partial x_j}\right] + P_k - \bar{\rho}(\epsilon + \epsilon_c) + \overline{p''d''} + L_k \quad (21\text{-}169)$$

and

$$\frac{\partial}{\partial t}(\bar{\rho}\,\epsilon) + \frac{\partial}{\partial x_j}(\bar{\rho}\,\bar{u}_j\,\epsilon) = \frac{\partial}{\partial x_j}\left[\left(\mu + \frac{\mu_t}{\sigma_\epsilon}\right)\frac{\partial \epsilon}{\partial x_j}\right] + c_{\epsilon 1}\, f_1\, P_k\frac{\epsilon}{k} - c_{\epsilon 2}\, f_2\,\bar{\rho}\frac{\epsilon^2}{k} + L_\epsilon \quad (21\text{-}170)$$

The term ϵ_c represents the contribution due to compressible dissipation, and the term $\overline{p''d''}$ represents the pressure dilatation term. They are given respectively by

$$\epsilon_c = \gamma_1\, M_t^2\, \epsilon \quad (21\text{-}171)$$

and

$$\overline{p''d''} = -\gamma_2\, P_k\, M_t^2 + \gamma_3\, \rho\epsilon\, M_t^2 \quad (21\text{-}172)$$

The following values based on DNS have been suggested.

$$\gamma_1 = 1.0\,, \qquad \gamma_2 = 0.4\,, \quad \text{and} \quad \gamma_3 = 0.2$$

The simplest approach to include axisymmetric correction is to modify the coefficient $c_{\epsilon 1}$ as

$$c_{\epsilon 1} = \begin{cases} 1.44 & \text{for planar 2-D flows} \\ 1.60 & \text{for axisymmetric 2-D flows} \end{cases}$$

21.4.3.1.3 Initial and boundary conditions: The initial and boundary conditions must be specified for both k and ϵ. The initial condition for the turbulent kinetic energy k is specified in terms of the freestream turbulence intensity T_i, as follows

$$k_\infty = 1.5(T_i\, U_\infty)^2$$

where $T_i = 0.0001$ to 10 percent of mean velocity. The turbulent viscosity is specified as

$$\mu_{t_\infty} = (0.1 \rightarrow 100)\mu_\infty$$

and, therefore,

$$\epsilon_\infty = \bar{\rho} c_\mu \frac{k_\infty^2}{\mu_{t_\infty}}$$

The boundary conditions are

(1) At the inflow, the freestream values are specified, that is, $\epsilon = \epsilon_\infty$ and $k = k_\infty$.

(2) At the solid surface, $k = 0$, $\mu_t = 0$, and $\epsilon = \epsilon_w = 0$. For some k-ϵ models, a value for ϵ_w is prescribed.

(3) At the outflow, extrapolation is used.

21.4.3.2 k-ω Two-Equation Turbulence Model: This two-equation model includes one equation for the turbulent kinetic energy k, as developed previously, and a second equation for the specific turbulent dissipation rate (or turbulent frequency) ω.

Recall that the development of the k-equation was based on the governing equations of fluid motion. A second approach for the development of a transport equation may be considered whereby the equation is constructed based on the known physical processes along with dimensional analysis. This is the approach taken to establish a transport equation for ω.

The concept of parameter ω was introduced by Kolmogorov, which he called *dissipation per unit turbulence kinetic energy.* Recall that the process of dissipation takes place at the level of smallest eddies, while the rate of dissipation is the rate of transfer of turbulence energy to the smallest eddies. Therefore, this rate is set by the properties of the large eddies. Perhaps the simplest physical interpretation of ω is that it represents the ratio of turbulent dissipation rate to the turbulence mixing energy or, alternatively, the rate of dissipation of turbulence per unit energy which in essence is an inverse time scale of the large eddies.

Now, to develop a transport equation for ω, several physical processes which are observed in fluids are utilized. These processes include unsteadiness, convection, diffusion, dissipation, and production. A combination of these physical processes, along with dimensional arguments, yield

$$\rho \frac{\partial \omega}{\partial t} + \rho u_j \frac{\partial \omega}{\partial x_j} = -\beta \rho \omega^2 + \frac{\partial}{\partial x_j}\left[\sigma \mu_t \frac{\partial \omega}{\partial x_j}\right] \tag{21-173}$$

where β and σ are coefficients to be specified.

In comparison to the k-equation, several observations can be made. First, note that a production term does not appear in Equation (21-173). Second, a molecular diffusion term is also absent from this equation. Inclusion of a molecular diffusion term is essential if one intends to integrate the equation through the viscous sublayer to the wall. Equation (21-173) is considered as the basic ω-equation upon which modifications are introduced.

As in the case of the k-ϵ model, there are several versions of the k-ω model. Perhaps one of the best known is the k-ω model of Wilcox [21-17], which is provided below.

The turbulent kinetic energy equation is given by

$$\bar{\rho}\frac{dk}{dt} = \frac{\partial}{\partial x_j}\left[(\mu + \sigma^* \mu_t)\frac{\partial k}{\partial x_j}\right] + P_k - \beta^* \rho \omega k \tag{21-174}$$

and the specific dissipation rate equation is

$$\bar{\rho}\frac{d\omega}{dt} = \frac{\partial}{\partial x_j}\left[(\mu + \sigma \mu_t)\frac{\partial \omega}{\partial x_j}\right] + \alpha\frac{\omega}{k}P_k - \beta \rho \omega^2 \tag{21-175}$$

The eddy viscosity is determined from

$$\mu_t = \rho\frac{k}{\omega} \tag{21-176}$$

and the auxiliary relations are

$$\epsilon = \beta^* \omega k \tag{21-177}$$

$$\ell = \frac{k^{\frac{1}{2}}}{\omega} \tag{21-178}$$

The constants used in Equations (21-174) and (21-175) are

$$\alpha = \frac{5}{9} \qquad \beta = \frac{3}{40} \qquad \beta^* = \frac{9}{100} \qquad \sigma = \frac{1}{2} \qquad \sigma^* = \frac{1}{2}$$

21.4.3.3 Combined k-ε/k-ω Two-equation Turbulence Model: The k-ω turbulence model performs well and, in fact, is superior to the k-ϵ model within the laminar sublayer and the logarithmic region of the boundary layer. However, the k-ω model has been shown to be influenced strongly by the specification of freestream value of ω outside the boundary layer. Therefore, the k-ω model does not appear to be an ideal model for applications in the wake region of the boundary layer. On the other hand, the k-ϵ model behaves superior to that of the k-ω model in the outer portion and wake regions of the boundary layer, but inferior in the inner region of the boundary layer.

To include the best features of each model, Menter [21-18] has combined different elements of the k-ϵ and k-ω models to form a new two-equation turbulence model. This model incorporates the k-ω model for the inner region of the boundary layer, and it switches to the k-ϵ model for the outer and wake regions of the boundary layer. Two versions of the model introduced by Menter are referred to as the *baseline (BSL) model* and, a modified version of the BSL model, the *shear-stress transport (SST) model*. It has been shown that the SST model performs well in the prediction of flows with adverse pressure gradient.

The baseline model is identical to the k-ω model given by Equations (21-174) and (21-175) in the sublayer and the log layer of the boundary layer, and it gradually switches to the k-ϵ model given by Equation (21-153) and (21-154) in the outer wake region. These equations are repeated here for convenience. The k-ω equations are

$$\frac{\partial}{\partial t}(\bar{\rho} k) + \frac{\partial}{\partial x_j}(\bar{\rho}\,\bar{u}_j\,k) = \frac{\partial}{\partial x_j}\left[(\mu + \sigma_{k1}\,\mu_t)\,\frac{\partial k}{\partial x_j}\right] + P_k - \beta^* \bar{\rho} k\omega \tag{21-179}$$

and

$$\frac{\partial}{\partial t}(\bar{\rho}\omega) + \frac{\partial}{\partial x_j}(\bar{\rho}\,\bar{u}_j\,\omega) = \frac{\partial}{\partial x_j}\left[(\mu + \sigma_{\omega 1}\,\mu_t)\,\frac{\partial \omega}{\partial x_j}\right] + \alpha_1 \frac{\omega}{k} P_k - \beta_1 \bar{\rho}\omega^2 \tag{21-180}$$

where

$$\mu_t = \rho \frac{k}{\omega} \tag{21-181}$$

and

$$\epsilon = \beta^* \omega\, k \tag{21-182}$$

Furthermore, the following are used

$$\sigma_k = \sigma^* \;, \qquad \sigma_{\omega 1} = \sigma \;, \qquad \beta_1 = \beta \;, \qquad \text{and} \quad \alpha_1 = \alpha$$

The k-ϵ equations are

$$\frac{\partial}{\partial t}(\bar{\rho} k) + \frac{\partial}{\partial x_j}(\bar{\rho}\bar{u}_j k) = \frac{\partial}{\partial x_j}\left[\left(\mu + \frac{\mu_t}{\sigma_k}\right)\frac{\partial k}{\partial x_j}\right] + P_k - \bar{\rho}\epsilon \tag{21-183}$$

and

$$\frac{\partial}{\partial t}(\bar{\rho}\epsilon) + \frac{\partial}{\partial x_j}(\bar{\rho}\,\bar{u}_j\,\epsilon) = \frac{\partial}{\partial x_j}\left[\left(\mu + \frac{\mu_t}{\sigma_\epsilon}\right)\frac{\partial \epsilon}{\partial x_j}\right] + c_{\epsilon 1}\,\frac{\epsilon}{k}\,P_k - c_{\epsilon 2}\,\bar{\rho}\,\frac{\epsilon^2}{k} \tag{21-184}$$

where

$$\mu_t = \rho\, c_\mu \frac{k^2}{\epsilon} \tag{21-185}$$

and

$$\omega = \frac{\epsilon}{c_\mu k} \tag{21-186}$$

Now, in order to combine the two sets of equations, the k-ϵ set is transformed into a k-ω formulation using relation (21-182). In the process of this transformation, two additional terms appear in the new ω-equation. One of these terms is a cross-diffusion term and the other occurs if σ_k and σ_ϵ are not equal. It has been shown that the second term is relatively small and does not affect significantly the solution. Therefore it is neglected. Subsequently, the transformed k-ϵ equations written in a k-ω formulation are written as

$$\frac{\partial}{\partial t}(\bar{\rho} k) + \frac{\partial}{\partial x_j}(\bar{\rho}\,\bar{u}_j\,k) = \frac{\partial}{\partial x_j}\left[(\mu + \sigma_{k2}\,\mu_t)\,\frac{\partial k}{\partial x_j}\right] + P_k - \beta^* \bar{\rho} k \omega \tag{21-187}$$

and

$$\frac{\partial}{\partial t}(\bar{\rho}\omega) + \frac{\partial}{\partial x_j}(\bar{\rho}\,\bar{u}_j\,\omega) = \frac{\partial}{\partial x_j}\left[(\mu + \sigma_{\omega 2}\,\mu_t)\frac{\partial \omega}{\partial x_j}\right]$$

$$+\,\alpha_2\,\frac{\omega}{k} P_k - \beta_2\,\bar{\rho}\,\omega^2 + 2\bar{\rho}\,\sigma_{\omega 2}\,\frac{1}{\omega}\,\frac{\partial k}{\partial x_j}\,\frac{\partial \omega}{\partial x_j} \tag{21-188}$$

The relations between the coefficients in the original k-ϵ equations and the transformed set are established as follows

$$\sigma_{k2} = \frac{1}{\sigma_k}\,, \qquad \sigma_{\omega 2} = \frac{1}{\sigma_\epsilon}\,, \qquad \beta_2 = \beta^*(c_{\epsilon 2} - 1)$$

$$\beta^* = c_\mu\,, \qquad \alpha_2 = (c_{\epsilon 1} - 1)$$

Now the two sets of equations are combined by the introduction of a blending function F. The k-ω set given by Equations (21-179) and (21-180) is multiplied by F, and the transofmred k-ϵ set given by Equations (21-187) and (21-188) is multiplied by $(1-F)$. They are subsequently added together. The blending function F is designed such that it will be equal to 1 in the vicinity of the wall region, thereby activating the k-ω model, and it will be zero away from the wall, thus activating the transformed k-ϵ model.

The resulting combined k-ω/k-ϵ two-equation model is given by

$$\frac{\partial}{\partial t}(\bar{\rho}\,k) + \frac{\partial}{\partial x_j}(\bar{\rho}\,\bar{u}_j\,k) = \frac{\partial}{\partial x_j}\left[(\mu + \sigma_k \mu_t)\frac{\partial k}{\partial x_j}\right] + P_k - \beta^* \rho \omega k \tag{21-189}$$

and

$$\frac{\partial}{\partial t}(\bar{\rho}\,\omega) + \frac{\partial}{\partial x_j}(\bar{\rho}\,\bar{u}_j\,\omega) = \frac{\partial}{\partial x_j}\left[(\mu + \sigma_\omega \mu_t)\frac{\partial \omega}{\partial x_j}\right] +$$

$$2(1 - F_1)\bar{\rho}\sigma_{\omega 2}\frac{1}{\omega}\,\frac{\partial k}{\partial x_j}\,\frac{\partial \omega}{\partial x_j} + \alpha\frac{\omega}{k}P_k - \beta\bar{\rho}\omega^2 \tag{21-190}$$

where the production of turbulence as defined previously is given by

$$P_k = \tau_{ij}\frac{\partial u_i}{\partial x_j} \tag{21-191}$$

and

$$\tau_{ij} = \mu_t\left(\frac{\partial u_i}{\partial x_j} + \frac{\partial u_j}{\partial x_i} - \frac{2}{3}\delta_{ij}\frac{\partial u_k}{\partial x_k}\right) - \frac{2}{3}\rho k\delta_{ij}$$

The constants appearing in Equations (21-189) and (21-290) are expressed in a general compact form by

$$\phi = F_1\phi_1 + (1 - F_1)\phi_2 \tag{21-192}$$

where ϕ_1 represents the constants associated with the k-ω model (when $F_1 = 1$), and ϕ_2 represents the constants associated with the k-ϵ model (when $F_1 = 0$).

The difference between the baseline model and the shear stress transport model is in the definition of turbulent viscosity and the specification of constants.

21.4.3.3.1 Baseline model: The turbulent viscosity for the baseline model is defined by

$$\mu_t = \bar{\rho}\frac{k}{\omega} \tag{21-193}$$

and the constants for ϕ are set as follows.

The values of the constants for set ϕ_1 are specified according to

$$\sigma_{k1} = 0.5 \ , \qquad \sigma_{\omega 1} = 0.5 \ , \qquad \beta_1 = 0.075$$

$$\beta^* = 0.09 \ , \qquad \kappa = 0.41$$

$$\alpha_1 = \frac{\beta_1}{\beta^*} - \sigma_{\omega 1}\kappa^2/\sqrt{\beta^*} \cong \frac{5}{9}$$

And the constants for ϕ_2 are

$$\sigma_{k2} = \frac{1}{\sigma_k} = 1.0 \ , \qquad \sigma_{\omega 2} = \frac{1}{\sigma_\epsilon} = 0.856 \ , \quad \beta_2 = 0.0828$$

$$\beta^* = 0.09 \ , \qquad \kappa = 0.41$$

$$\alpha_2 = \frac{\beta_2}{\beta^*} - \sigma_{\omega 2}\kappa^2/\sqrt{\beta^*}$$

In addition, the following definitions are used.

$$F_1 = \text{Tanh}\,(\text{arg}_1^4) \tag{21-194}$$

$$\text{arg}_1 = \min\left[\max\left(\frac{\sqrt{k}}{0.09\omega y}\ ,\ \frac{500\nu}{\omega y^2}\right)\ ,\ \frac{4\rho\sigma_{\omega 2}k}{CD_{k\omega}y^2}\right] \tag{21-195}$$

where y is the distance to the nearest surface, and $CD_{k\omega}$ is the positive portion of the cross-diffusion term

$$CD_{k\omega} = \max\left[2\rho\,\sigma_{\omega 2}\frac{1}{\omega}\frac{\partial k}{\partial x_i}\frac{\partial \omega}{\partial x_i}\ ,\ 10^{-20}\right] \tag{21-196}$$

Note that far away from solid surface as $y \rightarrow$ large, term $\arg_1$ approaches zero due to $1/y$ and $1/y^2$. Furthermore, the three arguments within $\arg_1$ represent the following:

1) The term $(\sqrt{k}/0.09\omega y)$ is the turbulence length scale divided by y. It is equal to 2.5 in the log layer of the boundary layer and approaches zero at the boundary-layer edge.

2) The term $(500\nu/\omega y^2)$ enforces F_1 to be one in the sublayer region and $1/y$ in the log layer. Recall that ω is proportional to $1/y^2$ near the surface, and it is proportional to $1/y$ in the log region. Therefore, $1/\omega y^2$ will be constant near the surface and will approach zero in the log layer.

3) The third term $(4\rho\sigma_{\omega 2}k/CD_{\omega k}y^2)$ is included to prevent solution dependency on the freestream. It can be shown that, as the boundary-layer edge is approached, $\arg_1$ as well as F_1 become zero and, therefore, the k-ϵ model is utilized in this region.

Transition to turbulence is activated by phasing in the production term, typically over a few grid points at a specified transition point. Note that the production term is turned off in the laminar portion of the flow.

21.4.3.3.2 Initial and boundary conditions: The following freestream conditions are recommended.

$$\omega_\infty = m\frac{u_\infty}{L}\ , \qquad \nu_{t_\infty} = 10^{-n}\nu_\infty\ , \qquad k_\infty = \nu_{t_\infty}\omega_\infty$$

where

$$1 \le m \le 10\ , \qquad 2 \le n \le 5\ ,$$

and L is a characteristic length of the problem, for example, the length of computational domain. At the solid surface, the boundary condition for ω is set according to

$$\omega = 10\frac{6\nu}{\beta_1(\Delta y_1)^2} \tag{21-197}$$

where Δy_1 is the distance of the first point away from the wall. This boundary condition is specified for a smooth wall, and Δy_1^+ must be less than about 3. At the outflow boundary, extrapolation is used.

21.4.3.3.3 Shear-stress transport model: The turbulent viscosity of the BSL model is modified based on the assumption that the turbulent shear stress is proportional to the turbulent kinetic energy in the logarithmic and wake regions of the turbulent boundary layer. The amount of computed turbulent viscosity is limited in order to satisfy the proportionality requirement. This limitation on the turbulent viscosity is imposed on by the turbulent kinetic energy, and thus the turbulent viscosity is defined as

$$\nu_t = \frac{a_1 k}{\max(a_1 \omega \, , \, \Omega F_2)} \tag{21-198}$$

where $a_1 = 0.31$, Ω is the absolute value of vorticity $\Omega = |\partial v/\partial x - \partial u/\partial y|$, and $F_2 = \text{Tanh}\,(\text{arg}_2^2)$ with

$$\text{arg}_2 = \max\left[2\frac{\sqrt{k}}{0.09\omega y}\, , \, \frac{500\nu}{\omega y^2}\right] \tag{21-199}$$

The constants for ϕ_1 and ϕ_2 are identical to the baseline model, except σ_{k1}, where

$$\sigma_{k1} = 0.85$$

21.4.3.3.4 Compressibility correction: Similar to that of the k-ϵ model, the compressibility correction term is included by introduction of turbulent Mach number. Thus, Equations (17-189) and (17-190) are formulated as

$$\begin{aligned} \frac{\partial}{\partial t}(\bar{\rho} k) + \frac{\partial}{\partial x_j}(\bar{\rho}\, \bar{u}_j\, k) &= \frac{\partial}{\partial x_j}\left[(\mu + \sigma_k\, \mu_t)\frac{\partial k}{\partial x_j}\right] + P_k \\ &\quad - \beta^* \bar{\rho}\, \omega\, k\left[1 + \alpha_1 M_t^2 (1 - F_1)\right] + (1 - F_1)\overline{p'' d''} \end{aligned} \tag{21-200}$$

and

$$\begin{aligned} \frac{\partial}{\partial t}(\bar{\rho}\,\omega) + \frac{\partial}{\partial x_j}(\bar{\rho}\, \bar{u}_j\, \omega) &= \frac{\partial}{\partial x_j}\left[(\mu + \sigma_\omega\, \mu_t)\frac{\partial \omega}{\partial x_j}\right] \\ &\quad + \alpha\frac{\omega}{k} P_k - \beta\, \bar{\rho}\, \omega^2 + 2(1 - F_1)\bar{\rho}\, \sigma_{\omega 2}\frac{1}{\omega}\frac{\partial k}{\partial x_j}\frac{\partial \omega}{\partial x_j} \\ &\quad + (1 - F_1)\beta^* \alpha_1\, M_t^2\, \bar{\rho}\, \omega^2 - \frac{1}{\nu_t}(1 - F_1)\overline{p'' d''} \end{aligned} \tag{21-201}$$

where the blending function F_1 is defined as in the original model.

21.4.3.3.5 Nondimensional form: The nondimensional variables defined previously are used to express Equations (21-189) and (21-190) in a nondimensional form. Again, the notation of asterisk has been dropped. However, the appearance of the

Reynolds number in the equation is a good indication of nondimensional equation. The nondimensional k-ϵ/k-ω two-equation model is written as

$$\rho\frac{dk}{dt} = \rho\left(\frac{\partial k}{\partial t} + u\frac{\partial k}{\partial x} + v\frac{\partial k}{\partial y}\right) = \frac{1}{Re_\infty}\frac{\partial}{\partial x}\left[(\mu + \sigma_k\mu_t)\frac{\partial k}{\partial x}\right]$$

$$+\frac{1}{Re_\infty}\frac{\partial}{\partial y}\left[(\mu + \sigma_k\,\mu_t)\frac{\partial k}{\partial y}\right] + P_k - \beta^*\rho\omega k \tag{21-202}$$

and

$$\rho\frac{d\omega}{dt} = \rho\left(\frac{\partial \omega}{\partial t} + u\frac{\partial \omega}{\partial x} + v\frac{\partial \omega}{\partial y}\right) = \frac{1}{Re_\infty}\frac{\partial}{\partial x}\left[(\mu + \sigma_\omega\,\mu_t)\frac{\partial \omega}{\partial x}\right]$$

$$+\frac{1}{Re_\infty}\frac{\partial}{\partial y}\left[(\mu + \sigma_\omega\,\mu_t)\frac{\partial \omega}{\partial y}\right] + 2\rho(1 - F_1)\sigma_{\omega 2}\frac{1}{\omega}\left[\frac{\partial k}{\partial x}\frac{\partial \omega}{\partial x} + \frac{\partial k}{\partial y}\frac{\partial \omega}{\partial y}\right]$$

$$+\alpha\frac{\omega}{k}P_k - \beta\rho\omega^2 \tag{21-203}$$

where the production term P_k is determined from

$$P_k = \tau_{ij}\frac{\partial u_i}{\partial x_j} = \left[\mu_t\left(\frac{\partial u_i}{\partial x_j} + \frac{\partial u_j}{\partial x_i} - \frac{2}{3}\delta_{ij}\frac{\partial u_k}{\partial x_k}\right) - \frac{2}{3}\rho k\delta_{ij}\right]\frac{\partial u_i}{\partial x_j} \tag{21-204}$$

21.4.3.3.6 k-ϵ/k-ω Turbulence model in computational space: Using the expressions for the Cartesian derivatives given by (21-88) and (21-89), the k-equation in the computational space becomes

$$\rho\frac{\partial k}{\partial t} = -\rho U\frac{\partial k}{\partial \xi} - \rho V\frac{\partial k}{\partial \eta} + \frac{1}{Re_\infty}\Bigg\{\xi_x\frac{\partial}{\partial \xi}\left[(MUS)\,(KX)\right] + \eta_x\frac{\partial}{\partial \eta}\left[(MUS)\,(KX)\right] +$$

$$\xi_y\frac{\partial}{\partial \xi}\left[(MUS)\,(KY)\right] + \eta_y\frac{\partial}{\partial \eta}\left[(MUS)\,(KY)\right]\Bigg\} + P_k - \beta^*\rho\omega k \tag{21-205}$$

where

$$MUS = \mu + \sigma_k\mu_t \tag{21-206}$$

$$KX = \xi_x\frac{\partial k}{\partial \xi} + \eta_x\frac{\partial k}{\partial \eta} \tag{21-207}$$

$$KY = \xi_y\frac{\partial k}{\partial \xi} + \eta_y\frac{\partial k}{\partial \eta} \tag{21-208}$$

Similarly, the ω-equation is given by

$$\rho\frac{\partial \omega}{\partial t} = -\rho U\frac{\partial \omega}{\partial \xi} - \rho V\frac{\partial \omega}{\partial \eta} + \frac{1}{Re_\infty}\left\{\xi_x\frac{\partial}{\partial \xi}\left[(MUR)(OX)\right]\right.$$

$$\left. + \eta_x\frac{\partial}{\partial \eta}\left[(MUR)(OX)\right] + \xi_y\frac{\partial}{\partial \xi}\left[(MUR)(OY)\right] + \eta_y\frac{\partial}{\partial \eta}\left[(MUR)(OY)\right]\right\}$$

$$+\left[2\rho(1-F_1)\sigma_{\omega 2}\frac{1}{\omega}\right]\left[(KX)(OX)+(KY)(OY)\right] - \beta\rho\omega^2$$

(21-209)

where

$$MUR = \mu + \sigma_\omega \mu_t \quad (21\text{-}210)$$

$$OX = \xi_x\frac{\partial \omega}{\partial \xi} + \eta_x\frac{\partial \omega}{\partial \eta} \quad (21\text{-}211)$$

$$OY = \xi_y\frac{\partial \omega}{\partial \xi} + \eta_y\frac{\partial \omega}{\partial \eta} \quad (21\text{-}212)$$

The production term P_k given by (21-204) is first expanded and subsequently transformed to the computational space as follows.

$$P_k = \tau_{ij}\frac{\partial u_i}{\partial x_j} = \tau_{xx}\frac{\partial u}{\partial x} + \tau_{xy}\left(\frac{\partial u}{\partial y} + \frac{\partial v}{\partial x}\right) + \tau_{yy}\frac{\partial v}{\partial y} \quad (21\text{-}213)$$

where

$$\tau_{xx} = \left(\frac{1}{Re_\infty}\right)\mu_t\left(\frac{4}{3}\frac{\partial u}{\partial x} - \frac{2}{3}\frac{\partial v}{\partial y}\right) - \frac{2}{3}\rho k \quad (21\text{-}214)$$

$$\tau_{xy} = \left(\frac{1}{Re_\infty}\right)\mu_t\left(\frac{\partial u}{\partial y} + \frac{\partial v}{\partial x}\right) \quad (21\text{-}215)$$

$$\tau_{yy} = \left(\frac{1}{Re_\infty}\right)\mu_t\left(\frac{4}{3}\frac{\partial v}{\partial y} - \frac{2}{3}\frac{\partial u}{\partial x}\right) - \frac{2}{3}\rho k \quad (21\text{-}216)$$

Utilizing the expressions (21-88) and (21-89), the shear stresses defined by (21-214) through (21-216) become

$$\tau_{xx} = \left(\frac{1}{Re_\infty}\right)\mu_t\left[\frac{4}{3}\left(\xi_x\frac{\partial u}{\partial \xi} + \eta_x\frac{\partial u}{\partial \eta}\right) - \frac{2}{3}\left(\xi_y\frac{\partial v}{\partial \xi} + \eta_y\frac{\partial v}{\partial \eta}\right)\right] - \frac{2}{3}\rho k = \tau_{\xi\xi} \quad (21\text{-}217)$$

$$\tau_{xy} = \left(\frac{1}{Re_\infty}\right)\mu_t\left[\xi_y\frac{\partial u}{\partial \xi} + \eta_y\frac{\partial u}{\partial \eta} + \xi_x\frac{\partial v}{\partial \xi} + \eta_x\frac{\partial v}{\partial \eta}\right] = \tau_{\xi\eta} \tag{21-218}$$

$$\tau_{yy} = \left(\frac{1}{Re_\infty}\right)\mu_t\left[\frac{4}{3}\left(\xi_y\frac{\partial v}{\partial \xi} + \eta_y\frac{\partial v}{\partial \eta}\right) - \frac{2}{3}\left(\xi_x\frac{\partial u}{\partial \xi} + \eta_x\frac{\partial u}{\partial \eta}\right)\right] - \frac{2}{3}\rho k = \tau_{\eta\eta} \tag{21-219}$$

Now, the production term becomes

$$\begin{aligned} P_k &= \tau_{xx}\frac{\partial u}{\partial x} + \tau_{xy}\left(\frac{\partial u}{\partial y} + \frac{\partial v}{\partial x}\right) + \tau_{yy}\frac{\partial u}{\partial y} \\ &= \tau_{\xi\xi}\left(\xi_x\frac{\partial u}{\partial \xi} + \eta_x\frac{\partial u}{\partial \eta}\right) + \tau_{\xi\eta}\left(\xi_y\frac{\partial u}{\partial \xi} + \eta_y\frac{\partial u}{\partial \eta}\right. \\ &\quad \left. + \xi_x\frac{\partial v}{\partial \xi} + \eta_x\frac{\partial v}{\partial \eta}\right) + \tau_{\eta\eta}\left(\xi_y\frac{\partial v}{\partial \xi} + \eta_y\frac{\partial v}{\partial \eta}\right) \\ &= \left(\xi_x\,\tau_{\xi\xi} + \xi_y\,\tau_{\xi\eta}\right)\frac{\partial u}{\partial \xi} + \left(\eta_x\,\tau_{\xi\xi} + \eta_y\tau_{\xi\eta}\right)\frac{\partial u}{\partial \eta} \\ &\quad + \left(\xi_x\,\tau_{\xi\eta} + \xi_y\,\tau_{\eta\eta}\right)\frac{\partial v}{\partial \xi} + \left(\eta_x\,\tau_{\xi\eta} + \eta_y\,\tau_{\eta\eta}\right)\frac{\partial v}{\partial \eta} \\ &= (P1)\frac{\partial u}{\partial \xi} + (P2)\frac{\partial u}{\partial \eta} + (P3)\frac{\partial v}{\partial \xi} + (P4)\frac{\partial v}{\partial \eta} \end{aligned} \tag{21-220}$$

where

$$P1 = \xi_x\,\tau_{\xi\xi} + \xi_y\,\tau_{\xi\eta} \tag{21-221}$$

$$P2 = \eta_x\,\tau_{\xi\xi} + \eta_y\,\tau_{\xi\eta} \tag{21-222}$$

$$P3 = \xi_x\,\tau_{\xi\eta} + \xi_y\,\tau_{\eta\eta} \tag{21-223}$$

$$P4 = \eta_x\,\tau_{\xi\eta} + \eta_y\,\tau_{\eta\eta} \tag{21-224}$$

21.5 Numerical Considerations

Numerical simulation of turbulent flow may be divided into three categories. The simplest and most practical approach in use at present is the solution of the time-averaged Navier-Stokes equations along with a turbulence model. The second level of sophistication includes the simulation of time-dependent, large-scale eddy motion. In this approach, the effects of smaller scales are included by turbulence models. The third category includes the direct numerical simulation of all important scales of turbulence.

The second and third categories are presently in the infancy stage. One of the difficulties associated with these methods is the limitation of present day computers. However, with the advancement in computer technology, large eddy and full turbulent simulations seem possible for practical problems within the next decade or so.

Currently, the system of equations such as (21-27) along with some turbulence models are solved for engineering applications. The solution schemes for the system of equations given by (21-27) have been discussed in the previous chapters. One important point to emphasize here is grid resolution. In order to accurately resolve the feature of turbulent flow near the surface, a grid point within the laminar sublayer must be included. Generally, this corresponds to a grid point within y^+ of 2.0.

The inclusion of a turbulence model into the system of governing equations may be accomplished either explicitly or implicitly. One may solve the governing equations [such as (21-27)] implicitly and use the turbulence quantities from the previous time-level, i.e., explicitly. On the other hand, the governing equations, along with turbulence equations such as (21-155) and (21-156), may be solved simultaneously, i.e., implicitly. Obviously, a fully implicit formulation will increase the size of the Jacobian matrices and will require major modification of the (laminar) Navier-Stokes code. On the other hand, explicit treatment of turbulence is simple and straightforward, and modification of an existing Navier-Stokes code to include turbulence can be accomplished with ease.

21.6 Finite Difference Formulations

The one-equation and two-equation turbulence models in different forms given in Sections 21.4.2 and 21.4.3 are parabolic equations. Therefore, various numerical methods discussed previously for the solution of parabolic equations can be used to develop solution algorithms.

To illustrate the development of the numerical schemes for the turbulence models, both explicit and implicit formulations are considered. However, the implicit formulations will be limited to the one-equation models. The turbulence models in nondimensional form and in the generalized curvilinear coordinate system are considered to provide the FDEs for general applications.

21.6.1 Baldwin-Barth One-Equation Turbulence Model

A first-order upwind approximation is used for the convective terms, and a second-order central difference approximation is used for the diffusion terms. The finite difference expression for each term is developed separately due to the complexity of the equation. The first-order upwind scheme can be easily extended to a second-order upwind scheme.

21.6.1.1 Implicit Formulation: The Baldwin-Barth one-equation turbulence model written as

$$\frac{\partial R}{\partial t} = M + P \tag{21-225}$$

is used to illustrate the development of an implicit scheme. Equation (21-225) is written as

$$\left(\frac{\partial R}{\partial t}\right)^{n+1} = M^{n+1} + P^{n+1} \tag{21-226}$$

Since the governing equation is nonlinear, a linearization procedure similar to that introduced previously, for example, by relation (11-81), is used to provide the following.

$$M^{n+1} = M^n + \frac{\partial M}{\partial R}\Delta R = M^n + \overline{M}^n \Delta R \tag{21-227}$$

and

$$P^{n+1} = P^n + \frac{\partial P}{\partial R}\Delta R = P^n + \overline{P}^n \Delta R \tag{21-228}$$

Now Equation (21-226) is written as

$$\frac{\Delta R}{\Delta t} - (\overline{M} + \overline{P})\Delta R = (M^n + P^n) \tag{21-229}$$

Recall that the term M can be decomposed into terms involving ξ or η derivatives as defined by (21-107) through (21-110). Thus, M is expressed as $M = M_\xi + M_\eta$. Now, Equation (21-229) is written as

$$\frac{\Delta R}{\Delta t} - (\overline{M}_\xi + \overline{M}_\eta + \overline{P})\,\Delta R = (M_\xi^n + M_\eta^n + P^n) \tag{21-230}$$

or

$$\left[1 - (\overline{M}_\xi + \overline{M}_\eta + \overline{P})\Delta t\right]\Delta R = (M_\xi^n + M_\eta^n + P^n)\Delta t \tag{21-231}$$

and decomposed as

$$\left[1 - \Delta t\overline{M}_\xi\right]\left[1 - \Delta t(\overline{M}_\eta + \overline{P})\right]\Delta R = (M_\xi^n + M_\eta^n + P^n)\Delta t \tag{21-232}$$

Equation (21-232) is solved sequentially be the following equations.

$$(1 - \Delta t\,\overline{M}_\xi)\Delta R^* = (M_\xi^n + M_\eta^n + P^n)\Delta t \tag{21-233}$$

and

$$\left[1 - \Delta t(\overline{M}_\eta + \overline{P})\right] \Delta R = \Delta R^* \tag{21-234}$$

The finite difference expressions for each one of the terms appearing in Equations (21-233) and (21-234) are developed separately.

The convective term M^1 given by (21-106) is approximated by a first-order upwind formulation as follows.

$$\begin{aligned} M^1 &= -U\frac{\partial R}{\partial \xi} - V\frac{\partial R}{\partial \eta} \\ &= -\left\{\frac{1}{2}(U_{i,j} + |U_{i,j}|)\left(\frac{R_{i,j} - R_{i-1,j}}{\Delta\xi}\right)\right. \\ &\quad + \frac{1}{2}(U_{i,j} - |U_{i,j}|)\left(\frac{R_{i+1,j} - R_{i,j}}{\Delta\xi}\right) \\ &\quad + \frac{1}{2}(V_{i,j} + |V_{i,j}|)\left(\frac{R_{i,j} - R_{i,j-1}}{\Delta\eta}\right) \\ &\quad \left. + \frac{1}{2}(V_{i,j} - |V_{i,j}|)\left(\frac{R_{i,j+1} - R_{i,j}}{\Delta\eta}\right)\right\} \end{aligned} \tag{21-235}$$

If a second-order approximation for the convective term is desired, then the formulation is as follows.

$$\begin{aligned} M^1 &= -U\frac{\partial R}{\partial \xi} - V\frac{\partial R}{\partial \eta} \\ &= -\left\{\frac{1}{2}(U_{i,j} + |U_{i,j}|)\left(\frac{3R_{i,j} - 4R_{i-1,j} + R_{i-2,j}}{2\Delta\xi}\right)\right. \\ &\quad + \frac{1}{2}(U_{i,j} - |U_{i,j}|)\left(\frac{-R_{i+2,j} + 4R_{i+1,j} - 3R_{i,j}}{2\Delta\xi}\right) \\ &\quad + \frac{1}{2}(V_{i,j} + |V_{i,j}|)\left(\frac{3R_{i,j} - 4R_{i,j-1} + R_{i,j-2}}{2\Delta\eta}\right) \\ &\quad \left. + \frac{1}{2}(V_{i,j} - |V_{i,j}|)\left(\frac{-R_{i,j+2} + 4R_{i,j+1} - 3R_{i,j}}{2\Delta\eta}\right)\right\} \end{aligned} \tag{21-236}$$

Since all the terms within M^2 given by (21-107) or (21-108) are similar, the development of the finite difference expression will be illustrated in detail for one term only. Subsequently, the conclusion will be extended to the remaining terms. Now, consider the term $\xi_x(\partial\alpha/\partial\xi)$ which is approximated by a second-order central difference expression as follows.

$$\xi_x \frac{\partial \alpha}{\partial \xi} \cong \xi_{x_{i,j}} \frac{\alpha_{i+\frac{1}{2},j} - \alpha_{i-\frac{1}{2},j}}{\Delta\xi} \tag{21-237}$$

Furthermore, recall that from (21-84)

$$\alpha = \nu_t \left(\xi_x \frac{\partial R}{\partial \xi} + \eta_x \frac{\partial R}{\partial \eta} \right) \tag{21-238}$$

The finite difference expression for α can be written as

$$\alpha_{i+\frac{1}{2},j} = \nu_{t_{i+\frac{1}{2},j}} \left[\xi_{x_{i+\frac{1}{2},j}} \left(\frac{R_{i+1,j} - R_{i,j}}{\Delta\xi} \right) + \eta_{x_{i+\frac{1}{2},j}} \frac{R_{i+\frac{1}{2},j+\frac{1}{2}} - R_{i+\frac{1}{2},j-\frac{1}{2}}}{\Delta\eta} \right] \tag{21-239}$$

and

$$\alpha_{i-\frac{1}{2},j} = \nu_{t_{i-\frac{1}{2},j}} \left[\xi_{x_{i-\frac{1}{2},j}} \left(\frac{R_{i,j} - R_{i-1,j}}{\Delta\xi} \right) + \eta_{x_{i-\frac{1}{2},j}} \frac{R_{i-\frac{1}{2},j+\frac{1}{2}} - R_{i-\frac{1}{2},j-\frac{1}{2}}}{\Delta\eta} \right] \tag{21-240}$$

Substitution of (21-239) and (21-240) into (21-237) yields

$$\begin{aligned} \xi_x \frac{\partial \alpha}{\partial \xi} &= \xi_{x_{i,j}} \frac{1}{\Delta\xi} \Bigg\{ \nu_{t_{i+\frac{1}{2},j}} \left[\xi_{x_{i+\frac{1}{2},j}} \left(\frac{R_{i+1,j} - R_{i,j}}{\Delta\xi} \right) + \eta_{x_{i+\frac{1}{2},j}} \left(\frac{R_{i+\frac{1}{2},j+\frac{1}{2}} - R_{i+\frac{1}{2},j-\frac{1}{2}}}{\Delta\eta} \right) \right] \\ &\quad - \nu_{t_{i-\frac{1}{2},j}} \left[\xi_{x_{i-\frac{1}{2},j}} \left(\frac{R_{i,j} - R_{i-1,j}}{\Delta\xi} \right) + \eta_{x_{i-\frac{1}{2},j}} \left(\frac{R_{i-\frac{1}{2},j+\frac{1}{2}} - R_{i-\frac{1}{2},j-\frac{1}{2}}}{\Delta\eta} \right) \right] \Bigg\} \end{aligned} \tag{21-241}$$

A similar expression is developed for $\xi_y(\partial\beta/\partial\xi)$ and is combined with $\xi_x(\partial\alpha/\partial\xi)$ to provide an approximation for M_ξ^2 as follows.

$$M_\xi^2 = -\frac{1}{Re_\infty} \frac{1}{\sigma_\epsilon} \left(\xi_x \frac{\partial \alpha}{\partial \xi} + \xi_y \frac{\partial \beta}{\partial \xi} \right) \tag{21-242}$$

where

$$\begin{aligned} \xi_x \frac{\partial \alpha}{\partial \xi} + \xi_y \frac{\partial \beta}{\partial \xi} &\cong \nu_{t_{i+\frac{1}{2},j}} \left\{ (\xi_{x_{i,j}} \xi_{x_{i+\frac{1}{2},j}} + \xi_{y_{i,j}} \xi_{y_{i+\frac{1}{2},j}}) \left[\frac{R_{i+1,j} - R_{i,j}}{(\Delta\xi)^2} \right] \right\} \\ &\quad - \nu_{t_{i-\frac{1}{2},j}} \left\{ (\xi_{x_{i,j}} \xi_{x_{i-\frac{1}{2},j}} + \xi_{y_{i,j}} \xi_{y_{i-\frac{1}{2},j}}) \left[\frac{R_{i,j} - R_{i-1,j}}{(\Delta\xi)^2} \right] \right\} \end{aligned}$$

$$+\nu_{t_{i+\frac{1}{2},j}}\left\{\left(\xi_{x_{i,j}}\,\eta_{x_{i+\frac{1}{2},j}}+\xi_{y_{i,j}}\,\eta_{y_{i+\frac{1}{2},j}}\right)\left[\frac{R_{i+\frac{1}{2},j+\frac{1}{2}}-R_{i+\frac{1}{2},j-\frac{1}{2}}}{\Delta\xi\Delta\eta}\right]\right\}$$

$$-\nu_{t_{i-\frac{1}{2},j}}\left\{\left(\xi_{x_{i,j}}\,\eta_{x_{i-\frac{1}{2},j}}+\xi_{y_{i,j}}\,\eta_{y_{i-\frac{1}{2},j}}\right)\left[\frac{R_{i-\frac{1}{2},j+\frac{1}{2}}-R_{i-\frac{1}{2},j-\frac{1}{2}}}{\Delta\xi\Delta\eta}\right]\right\}$$

(21-243)

Similarly for M_η^2,

$$\eta_x\frac{\partial\alpha}{\partial\eta}+\eta_y\frac{\partial\beta}{\partial\eta}\cong\nu_{t_{i,j+\frac{1}{2}}}\left\{\left(\eta_{x_{i,j}}\eta_{x_{i,j+\frac{1}{2}}}+\eta_{y_{i,j}}\eta_{y_{i,j+\frac{1}{2}}}\right)\left[\frac{R_{i,j+1}-R_{i,j}}{(\Delta\eta)^2}\right]\right\}$$

$$-\nu_{t_{i,j-\frac{1}{2}}}\left\{\left(\eta_{x_{i,j}}\eta_{x_{i,j-\frac{1}{2}}}+\eta_{y_{i,j}}\eta_{y_{i,j-\frac{1}{2}}}\right)\left[\frac{R_{i,j}-R_{i,j-1}}{(\Delta\eta)^2}\right]\right\}$$

$$+\nu_{t_{i,j+\frac{1}{2}}}\left\{\left(\eta_{x_{i,j}}\xi_{x_{i,j+\frac{1}{2}}}+\eta_{y_{i,j}}\xi_{y_{i,j+\frac{1}{2}}}\right)\left[\frac{R_{i+\frac{1}{2},j+\frac{1}{2}}-R_{i-\frac{1}{2},j+\frac{1}{2}}}{\Delta\xi\Delta\eta}\right]\right\}$$

$$-\nu_{t_{i,j-\frac{1}{2}}}\left\{\left(\eta_{x_{i,j}}\xi_{x_{i,j-\frac{1}{2}}}+\eta_{y_{i,j}}\xi_{y_{i,j-\frac{1}{2}}}\right)\left[\frac{R_{i+\frac{1}{2},j-\frac{1}{2}}-R_{i-\frac{1}{2},j-\frac{1}{2}}}{\Delta\xi\Delta\eta}\right]\right\} \quad (21\text{-}244)$$

The terms in M^3 are also similar to that of M^2. Recall that

$$M_\xi^3 = \frac{2}{Re_\infty}\left(\nu+\frac{\nu_t}{\sigma_\epsilon}\right)\left(\xi_x\frac{\partial A}{\partial\xi}+\xi_y\frac{\partial B}{\partial\xi}\right) \quad (21\text{-}245)$$

$$M_\eta^3 = \frac{2}{Re_\infty}\left(\nu+\frac{\nu_t}{\sigma_\epsilon}\right)\left(\eta_x\frac{\partial A}{\partial\eta}+\eta_y\frac{\partial B}{\partial\eta}\right) \quad (21\text{-}246)$$

and

$$A = \xi_x\frac{\partial R}{\partial\xi}+\eta_x\frac{\partial R}{\partial\eta} \quad (21\text{-}247)$$

$$B = \xi_y\frac{\partial R}{\partial\xi}+\eta_y\frac{\partial R}{\partial\eta} \quad (21\text{-}248)$$

Now the finite difference approximation for M_ξ^3 following (21-243) can be expressed as

$$M_\xi^3 = \frac{2}{Re_\infty}\left(\nu+\frac{\nu_t}{\sigma_\epsilon}\right)_{i,j}\left\{\left(\xi_{x_{i,j}}\,\xi_{x_{i+\frac{1}{2},j}}+\xi_{y_{i,j}}\,\xi_{y_{i+\frac{1}{2},j}}\right)\left[\frac{R_{i+1,j}-R_{i,j}}{(\Delta\xi)^2}\right]\right.$$

$$-\left(\xi_{x_{i,j}}\,\xi_{x_{i-\frac{1}{2},j}}+\xi_{y_{i,j}}\,\xi_{y_{i-\frac{1}{2},j}}\right)\left[\frac{R_{i,j}-R_{i-1,j}}{(\Delta\xi)^2}\right]$$

$$+\left(\xi_{x_{i,j}}\eta_{x_{i+\frac{1}{2},j}}+\xi_{y_{i,j}}\eta_{y_{i+\frac{1}{2},j}}\right)\left[\frac{R_{i+\frac{1}{2},j+\frac{1}{2}}-R_{i+\frac{1}{2},j-\frac{1}{2}}}{\Delta\xi\Delta\eta}\right]$$

$$-\left(\xi_{x_{i,j}}\eta_{x_{i-\frac{1}{2},j}}+\xi_{y_{i,j}}\eta_{y_{i-\frac{1}{2},j}}\right)\left[\frac{R_{i-\frac{1}{2},j+\frac{1}{2}}-R_{i-\frac{1}{2},j-\frac{1}{2}}}{\Delta\xi\Delta\eta}\right]\Bigg\} \quad (21\text{-}249)$$

Similarly, following (21-244)

$$M_\eta^3 = \frac{2}{Re_\infty}\left(\nu+\frac{\nu_t}{\sigma_\epsilon}\right)_{i,j}\left\{\left(\eta_{x_{i,j}}\eta_{x_{i,j+\frac{1}{2}}}+\eta_{y_{i,j}}\eta_{y_{i,j+\frac{1}{2}}}\right)\left[\frac{R_{i,j+1}-R_{i,j}}{(\Delta\eta)^2}\right]\right.$$

$$-\left(\eta_{x_{i,j}}\eta_{x_{i,j-\frac{1}{2}}}+\eta_{y_{i,j}}\eta_{y_{i,j-\frac{1}{2}}}\right)\left[\frac{R_{i,j}-R_{i,j-1}}{(\Delta\eta)^2}\right]$$

$$+\left(\eta_{x_{i,j}}\xi_{x_{i,j+\frac{1}{2}}}+\eta_{y_{i,j}}\xi_{y_{i,j+\frac{1}{2}}}\right)\left[\frac{R_{i+\frac{1}{2},j+\frac{1}{2}}-R_{i-\frac{1}{2},j+\frac{1}{2}}}{\Delta\xi\Delta\eta}\right]$$

$$-\left(\eta_{x_{i,j}}\xi_{x_{i,j-\frac{1}{2}}}+\eta_{y_{i,j}}\xi_{y_{i,j-\frac{1}{2}}}\right)\left[\frac{R_{i+\frac{1}{2},j-\frac{1}{2}}-R_{i-\frac{1}{2},j-\frac{1}{2}}}{\Delta\xi\Delta\eta}\right]\Bigg\} \quad (21\text{-}250)$$

At this point, the finite difference approximation of the terms in Equations (21-233) and (21-234) have been identified. However, the bar quantities such as $\overline{M}$ require some deliberation.

The procedure for obtaining these quantities is illustrated for $\overline{M}_\xi^1$ in detail, and the remaining terms can be obtained in a similar fashion.

Recall that M_ξ^1 is approximated by (21-235) when first-order approximation is used. Thus,

$$M_\xi^1 = -U\frac{\partial R}{\partial\xi} = -\left\{\frac{1}{2}(U_{i,j}+|U_{i,j}|)\left(\frac{R_{i,j}-R_{i-1,j}}{\Delta\xi}\right)\right.$$

$$\left.+\frac{1}{2}(U_{i,j}-|U_{i,j}|)\left(\frac{R_{i+1,j}-R_{i,j}}{\Delta\xi}\right)\right\} \quad (21\text{-}251)$$

Note that, as before, when second-order approximation is used, that is, Equation (21-236), the terms at grid points $i-2$ and $i+2$ are taken to the right-hand side of the equation and treated explicitly. Now from Equation (21-251) expressions for $\overline{M}_\xi^1$ are developed as follow.

$$\overline{M}_\xi^1\Big|_{i-1,j} = \frac{\partial M_\xi^1}{\partial R}\Big|_{i-1,j} = \left\{\frac{1}{2}(U_{i,j}+|U_{i,j}|)\left(\frac{1}{\Delta\xi}\right)\right\}$$

$$\overline{M}_\xi^1\Big|_{i,j} = \frac{\partial M_\xi^1}{\partial R}\Big|_{i,j} = -\left\{\frac{1}{2}(U_{i,j}+|U_{i,j}|)\left(\frac{1}{\Delta\xi}\right)\right.$$

$$+\frac{1}{2}(U_{i,j} - |U_{i,j}|)\left(\frac{-1}{\Delta\xi}\right)\Bigg\}$$

$$= -\left\{|U_{i,j}|\left(\frac{1}{\Delta\xi}\right)\right\}$$

$$\overline{M}^1_\xi\Big|_{i+1,j} = \frac{\partial M^1_\xi}{\partial R}\Big|_{i+1,j} = -\left\{\frac{1}{2}(U_{i,j} - |U_{i,j}|)\left(\frac{1}{\Delta\xi}\right)\right\}$$

Now, with all the terms in Equations (21-233) and (21-234) identified, the two tridiagonal systems of equations are solved sequentially by any tridiagonal solver, for example, by the scheme introduced in Appendix B.

21.6.1.2 Explicit Formulation: The Baldwin-Barth turbulence model given by (21-99) is considered to develop an explicit scheme. Again, as in the implicit formulation, an upwind approximation is used for the convective term which could be either first-order or second-order, and a central difference approximation of second-order is used for the diffusion terms. In the following formulation, a first-order upwind scheme is used. However, as seen in the previous section, extension to second-order is simple and straightforward.

$$R^{n+1} = R^n - \Delta t\Bigg\{\frac{1}{2}(U_{i,j} + |U_{i,j}|)\left(\frac{R_{i,j} - R_{i-1,j}}{\Delta\xi}\right) +$$

$$+\frac{1}{2}(U_{i,j} - |U_{i,j}|)\left(\frac{R_{i+1,j} - R_{i,j}}{\Delta\xi}\right)$$

$$+\frac{1}{2}(V_{i,j} + |V_{i,j}|)\left(\frac{R_{i,j} - R_{i,j-1}}{\Delta\eta}\right)$$

$$+\frac{1}{2}(V_{i,j} - |V_{i,j}|)\left(\frac{R_{i,j+1} - R_{i,j}}{\Delta\eta}\right)\Bigg\}$$

$$-\Delta t\left(\frac{1}{Re_\infty}\right)\left(\frac{1}{\sigma_\epsilon}\right)\left[\nu_{t_{i+\frac{1}{2},j}}\left\{(\xi_{x_{i,j}}\xi_{x_{i+\frac{1}{2},j}} + \xi_{y_{i,j}}\xi_{y_{i+\frac{1}{2},j}})\left[\frac{R_{i+1,j} - R_{i,j}}{(\Delta\xi)^2}\right]\right\}\right.$$

$$-\nu_{t_{i-\frac{1}{2},j}}\left\{(\xi_{x_{i,j}}\xi_{x_{i-\frac{1}{2},j}} + \xi_{y_{i,j}}\xi_{y_{i-\frac{1}{2},j}})\left[\frac{R_{i,j} - R_{i-1,j}}{(\Delta\xi)^2}\right]\right\}$$

$$+\nu_{t_{i+\frac{1}{2},j}}\left\{(\xi_{x_{i,j}}\eta_{x_{i+\frac{1}{2},j}} + \xi_{y_{i,j}}\eta_{y_{i+\frac{1}{2},j}})\left[\frac{R_{i+\frac{1}{2},j+\frac{1}{2}} - R_{i+\frac{1}{2},j-\frac{1}{2}}}{\Delta\xi\Delta\eta}\right]\right\}$$

$$- \nu_{t_{i-\frac{1}{2},j}} \left\{ \left(\xi_{x_{i,j}} \eta_{x_{i-\frac{1}{2},j}} + \xi_{y_{i,j}} \eta_{y_{i-\frac{1}{2},j}} \right) \left[\frac{R_{i-\frac{1}{2},j+\frac{1}{2}} - R_{i-\frac{1}{2},j-\frac{1}{2}}}{\Delta\xi\Delta\eta} \right] \right\}$$

$$+ \nu_{t_{i,j+\frac{1}{2}}} \left\{ \left(\eta_{x_{i,j}} \eta_{x_{i,j+\frac{1}{2}}} + \eta_{y_{i,j}} \eta_{y_{i,j+\frac{1}{2}}} \right) \left[\frac{R_{i,j+1} - R_{i,j}}{(\Delta\eta)^2} \right] \right\}$$

$$- \nu_{t_{i,j-\frac{1}{2}}} \left\{ \left(\eta_{x_{i,j}} \eta_{x_{i,j-\frac{1}{2}}} + \eta_{y_{i,j}} \eta_{y_{i,j-\frac{1}{2}}} \right) \left[\frac{R_{i,j} - R_{i,j-1}}{(\Delta\eta)^2} \right] \right\}$$

$$+ \nu_{t_{i,j+\frac{1}{2}}} \left\{ \left(\eta_{x_{i,j}} \xi_{x_{i,j+\frac{1}{2}}} + \eta_{y_{i,j}} \xi_{y_{i,j+\frac{1}{2}}} \right) \left[\frac{R_{i+\frac{1}{2},j+\frac{1}{2}} - R_{i-\frac{1}{2},j+\frac{1}{2}}}{\Delta\xi\Delta\eta} \right] \right\}$$

$$- \nu_{t_{i,j-\frac{1}{2}}} \left\{ \left(\eta_{x_{i,j}} \xi_{x_{i,j-\frac{1}{2}}} + \eta_{y_{i,j}} \xi_{y_{i,j-\frac{1}{2}}} \right) \left[\frac{R_{i+\frac{1}{2},j-\frac{1}{2}} - R_{i-\frac{1}{2},j-\frac{1}{2}}}{\Delta\xi\Delta\eta} \right] \right\} \Bigg]$$

$$+ \Delta t \left(\frac{2}{Re_\infty} \right) \left(\nu + \frac{\nu_t}{\sigma_\epsilon} \right)_{i,j} \Bigg[\left(\xi_{x_{i,j}} \xi_{x_{i+\frac{1}{2},j}} + \xi_{y_{i,j}} \xi_{y_{i+\frac{1}{2},j}} \right) \left[\frac{R_{i+1,j} - R_{i,j}}{(\Delta\xi)^2} \right]$$

$$- \left(\xi_{x_{i,j}} \xi_{x_{i-\frac{1}{2},j}} + \xi_{y_{i,j}} \xi_{y_{i-\frac{1}{2},j}} \right) \left[\frac{R_{i,j} - R_{i-1,j}}{(\Delta\xi)^2} \right]$$

$$+ \left(\xi_{x_{i,j}} \eta_{x_{i+\frac{1}{2},j}} + \xi_{y_{i,j}} \eta_{y_{i+\frac{1}{2},j}} \right) \left[\frac{R_{i+\frac{1}{2},j+\frac{1}{2}} - R_{i+\frac{1}{2},j-\frac{1}{2}}}{\Delta\xi\Delta\eta} \right]$$

$$- \left(\xi_{x_{i,j}} \eta_{x_{i-\frac{1}{2},j}} + \xi_{y_{i,j}} \eta_{y_{i-\frac{1}{2},j}} \right) \left[\frac{R_{i-\frac{1}{2},j+\frac{1}{2}} - R_{i-\frac{1}{2},j-\frac{1}{2}}}{\Delta\xi\Delta\eta} \right]$$

$$+ \left(\eta_{x_{i,j}} \eta_{x_{i,j+\frac{1}{2}}} + \eta_{y_{i,j}} \eta_{y_{i,j+\frac{1}{2}}} \right) \left[\frac{R_{i,j+1} - R_{i,j}}{(\Delta\eta)^2} \right]$$

$$- \left(\eta_{x_{i,j}} \eta_{x_{i,j-\frac{1}{2}}} + \eta_{y_{i,j}} \eta_{y_{i,j-\frac{1}{2}}} \right) \left[\frac{R_{i,j} - R_{i,j-1}}{(\Delta\eta)^2} \right]$$

$$+ \left(\eta_{x_{i,j}} \xi_{x_{i,j+\frac{1}{2}}} + \eta_{y_{i,j}} \xi_{y_{i,j+\frac{1}{2}}} \right) \left[\frac{R_{i+\frac{1}{2},j+\frac{1}{2}} - R_{i-\frac{1}{2},j+\frac{1}{2}}}{\Delta\xi\Delta\eta} \right]$$

$$- \left(\eta_{x_{i,j}} \xi_{x_{i,j-\frac{1}{2}}} + \eta_{y_{i,j}} \xi_{y_{i,j-\frac{1}{2}}} \right) \left[\frac{R_{i+\frac{1}{2},j-\frac{1}{2}} - R_{i-\frac{1}{2},j-\frac{1}{2}}}{\Delta\xi\Delta\eta} \right] \Bigg]$$

$$+ \Delta t \left(c_{\epsilon 2} f_{2_{i,j}} - c_{\epsilon 1} \right) \sqrt{c_\mu D_{1_{i,j}} D_{2_{i,j}}}\, S_{i,j} R_{i,j} \tag{21-252}$$

where

$$\begin{aligned}
S_{i,j}^2 &= a_{1_{i,j}}\left(\frac{u_{i+1,j}-u_{i-1,j}}{2\Delta\xi}\right)^2 + b_{1_{i,j}}\left(\frac{u_{i,j+1}-u_{i,j-1}}{2\Delta\eta}\right)^2 \\
&+ a_{2_{i,j}}\left(\frac{v_{i+1,j}-v_{i-1,j}}{2\Delta\xi}\right)^2 + b_{2_{i,j}}\left(\frac{v_{i,j+1}-v_{i,j-1}}{2\Delta\eta}\right)^2 \\
&+ 2c_{1_{i,j}}\left(\frac{u_{i+1,j}-u_{i-1,j}}{2\Delta\xi}\right)\left(\frac{u_{i,j+1}-u_{i,j-1}}{2\Delta\eta}\right) + 2c_{2_{i,j}}\left(\frac{v_{i+1,j}-v_{i-1,j}}{2\Delta\xi}\right)\left(\frac{v_{i,j+1}-v_{i,j-1}}{2\Delta\eta}\right) \\
&+ 2a_{3_{i,j}}\left(\frac{u_{i+1,j}-u_{i-1,j}}{2\Delta\xi}\right)\left(\frac{v_{i+1,j}-v_{i-1,j}}{2\Delta\xi}\right) + 2b_{3_{i,j}}\left(\frac{u_{i,j+1}-u_{i,j-1}}{2\Delta\eta}\right)\left(\frac{v_{i,j+1}-v_{i,j-1}}{2\Delta\eta}\right) \\
&+ 2c_{3_{i,j}}\left(\frac{u_{i+1,j}-u_{i-1,j}}{2\Delta\xi}\right)\left(\frac{v_{i,j+1}-v_{i,j-1}}{2\Delta\eta}\right) + 2c_{4_{i,j}}\left(\frac{u_{i,j+1}-u_{i,j-1}}{2\Delta\eta}\right)\left(\frac{v_{i+1,j}-v_{i-1,j}}{2\Delta\xi}\right)
\end{aligned} \tag{21-253}$$

21.6.2 Spalart-Allmaras One-Equation Turbulence Model

As in the previous section, an upwind approximation is used for the convective terms, and central difference approximation is used for the viscous terms. Since the approach is similar to that of the previous section, the final form of the finite difference expression and/or equations are provided in the following subsections.

21.6.2.1 Implicit Formulation: The convective term can be approximated by either a first-order or a second-order upwind formulation. The first-order approximation yields

$$\begin{aligned}
M^1 &= -\left(U\frac{\partial\overline{\nu}}{\partial\xi} + V\frac{\partial\overline{\nu}}{\partial\eta}\right) \\
&= -\left\{\frac{1}{2}(U_{i,j}+|U_{i,j}|)\left(\frac{\overline{\nu}_{i,j}-\overline{\nu}_{i-1,j}}{\Delta\xi}\right)\right. \\
&+ \frac{1}{2}(U_{i,j}-|U_{i,j}|)\left(\frac{\overline{\nu}_{i+1,j}-\overline{\nu}_{i,j}}{\Delta\xi}\right) \\
&+ \frac{1}{2}(V_{i,j}+|V_{i,j}|)\left(\frac{\overline{\nu}_{i,j}-\overline{\nu}_{i,j-1}}{\Delta\eta}\right)
\end{aligned}$$

$$+ \frac{1}{2}(V_{i,j} - |V_{i,j}|)\left(\frac{\overline{\nu}_{i,j+1} - \overline{\nu}_{i,j}}{\Delta\eta}\right)\Bigg\} \tag{21-254}$$

The second-order approximation provides

$$\begin{aligned} M^1 &= -\Bigg\{\frac{1}{2}(U_{i,j} + |U_{i,j}|)\left(\frac{3\overline{\nu}_{i,j} - 4\overline{\nu}_{i-1,j} + \overline{\nu}_{i-2,j}}{2\Delta\xi}\right) \\ &= +\frac{1}{2}(U_{i,j} - |U_{i,j}|)\left(\frac{-\overline{\nu}_{i+2,j} + 4\overline{\nu}_{i+1,j} - 3\overline{\nu}_{i,j}}{2\Delta\xi}\right) \\ &\quad +\frac{1}{2}(V_{i,j} + |V_{i,j}|)\left(\frac{3\overline{\nu}_{i,j} - 4\overline{\nu}_{i,j-1} + \overline{\nu}_{i,j-2}}{2\Delta\eta}\right) \\ &\quad +\frac{1}{2}(V_{i,j} - |V_{i,j}|)\left(\frac{-\overline{\nu}_{i,j+2} + 4\overline{\nu}_{i,j+1} - 3\overline{\nu}_{i,j}}{2\Delta\eta}\right)\Bigg\} \end{aligned} \tag{21-255}$$

The terms M^2 and M^3 given by (21-144) through (21-147) are approximated following the procedure illustrated in the previous section.

$$M_\xi^2 = \left(\frac{1}{Re_\infty}\right)\left(\frac{1+c_{b2}}{\sigma}\right)\left(\xi_x\frac{\partial\alpha}{\partial\xi} + \xi_y\frac{\partial\beta}{\partial\xi}\right)$$

where

$$\begin{aligned} \xi_x\frac{\partial\alpha}{\partial\xi} + \xi_y\frac{\partial\beta}{\partial\xi} &\cong (\nu+\overline{\nu})_{i+\frac{1}{2},j}\left\{\left(\xi_{x_{i,j}}\xi_{x_{i+\frac{1}{2},j}} + \xi_{y_{i,j}}\xi_{y_{i+\frac{1}{2},j}}\right)\left[\frac{\overline{\nu}_{i+1,j} - \overline{\nu}_{i,j}}{(\Delta\xi)^2}\right]\right\} \\ &\quad - (\nu+\overline{\nu})_{i-\frac{1}{2},j}\left\{\left(\xi_{x_{i,j}}\xi_{x_{i-\frac{1}{2},j}} + \xi_{y_{i,j}}\xi_{y_{i-\frac{1}{2},j}}\right)\left[\frac{\overline{\nu}_{i,j} - \overline{\nu}_{i-1,j}}{(\Delta\xi)^2}\right]\right\} \\ &\quad + (\nu+\overline{\nu})_{i+\frac{1}{2},j}\left\{\left(\xi_{x_{i,j}}\eta_{x_{i+\frac{1}{2},j}} + \xi_{y_{i,j}}\eta_{y_{i+\frac{1}{2},j}}\right)\left[\frac{\overline{\nu}_{i+\frac{1}{2},j+\frac{1}{2}} - \overline{\nu}_{i+\frac{1}{2},j-\frac{1}{2}}}{\Delta\xi\Delta\eta}\right]\right\} \\ &\quad - (\nu+\overline{\nu})_{i-\frac{1}{2},j}\left\{\left(\xi_{x_{i,j}}\eta_{x_{i-\frac{1}{2},j}} + \xi_{y_{i,j}}\eta_{y_{i-\frac{1}{2},j}}\right)\left[\frac{\overline{\nu}_{i-\frac{1}{2},j+\frac{1}{2}} - \overline{\nu}_{i-\frac{1}{2},j-\frac{1}{2}}}{\Delta\xi\Delta\eta}\right]\right\} \end{aligned} \tag{21-256}$$

and

$$M_\eta^2 = \left(\frac{1}{Re_\infty}\right)\left(\frac{1+c_{b2}}{\sigma}\right)\left(\eta_x\frac{\partial\alpha}{\partial\eta} + \eta_y\frac{\partial\beta}{\partial\eta}\right)$$

where

$$\eta_x\frac{\partial\alpha}{\partial\eta} + \eta_y\frac{\partial\beta}{\partial\eta} \cong (\nu+\overline{\nu})_{i,j+\frac{1}{2}}\left\{\left(\eta_{x_{i,j}}\eta_{x_{i,j+\frac{1}{2}}} + \eta_{y_{i,j}}\eta_{y_{i,j+\frac{1}{2}}}\right)\left[\frac{\overline{\nu}_{i,j+1} - \overline{\nu}_{i,j}}{(\Delta\eta)^2}\right]\right\}$$

$$- (\nu + \overline{\nu})_{i,j-\frac{1}{2}} \left\{ (\eta_{x_{i,j}} \eta_{x_{i,j-\frac{1}{2}}} + \eta_{y_{i,j}} \eta_{y_{i,j-\frac{1}{2}}}) \left[\frac{\overline{\nu}_{i,j} - \overline{\nu}_{i,j-1}}{(\Delta\eta)^2} \right] \right\}$$

$$+ (\nu + \overline{\nu})_{i,j+\frac{1}{2}} \left\{ (\eta_{x_{i,j}} \xi_{x_{i,j+\frac{1}{2}}} + \eta_{y_{i,j}} \xi_{y_{i,j+\frac{1}{2}}}) \left[\frac{\overline{\nu}_{i+\frac{1}{2},j+\frac{1}{2}} - \overline{\nu}_{i-\frac{1}{2},j+\frac{1}{2}}}{\Delta\xi\Delta\eta} \right] \right\}$$

$$- (\nu + \overline{\nu})_{i,j-\frac{1}{2}} \left\{ (\eta_{x_{i,j}} \xi_{x_{i,j-\frac{1}{2}}} + \eta_{y_{i,j}} \xi_{y_{i,j-\frac{1}{2}}}) \left[\frac{\overline{\nu}_{i+\frac{1}{2},j-\frac{1}{2}} - \overline{\nu}_{i-\frac{1}{2},j-\frac{1}{2}}}{\Delta\xi\Delta\eta} \right] \right\} \quad (21\text{-}257)$$

and

$$M_\xi^3 = -\left(\frac{1}{Re_\infty}\right) \frac{c_{b2}}{\sigma} (\nu + \overline{\nu}) \left(\xi_x \frac{\partial A}{\partial \xi} + \xi_y \frac{\partial B}{\partial \xi} \right)$$

where

$$- (\nu + \overline{\nu}) \left(\xi_x \frac{\partial A}{\partial \xi} + \xi_y \frac{\partial B}{\partial \xi} \right) \cong$$

$$- (\nu + \overline{\nu})_{i,j} \left\{ (\xi_{x_{i,j}} \, \xi_{x_{i+\frac{1}{2},j}} + \xi_{y_{i,j}} \, \xi_{y_{i+\frac{1}{2},j}}) \left[\frac{\overline{\nu}_{i+1,j} - \overline{\nu}_{i,j}}{(\Delta\xi)^2} \right] \right\}$$

$$+ (\nu + \overline{\nu})_{i,j} \left\{ (\xi_{x_{i,j}} \, \xi_{x_{i-\frac{1}{2},j}} + \xi_{y_{i,j}} \, \xi_{y_{i-\frac{1}{2},j}}) \left[\frac{\overline{\nu}_{i,j} - \overline{\nu}_{i-1,j}}{(\Delta\xi)^2} \right] \right\}$$

$$- (\nu + \overline{\nu})_{i,j} \left\{ (\xi_{x_{i,j}} \, \eta_{x_{i+\frac{1}{2},j}} + \xi_{y_{i,j}} \, \eta_{y_{i+\frac{1}{2},j}}) \left[\frac{\overline{\nu}_{i+\frac{1}{2},j+\frac{1}{2}} - \overline{\nu}_{i+\frac{1}{2},j-\frac{1}{2}}}{\Delta\xi\Delta\eta} \right] \right\}$$

$$+ (\nu + \overline{\nu})_{i,j} \left\{ (\xi_{x_{i,j}} \, \eta_{x_{i-\frac{1}{2},j}} + \xi_{y_{i,j}} \, \eta_{y_{i-\frac{1}{2},j}}) \left[\frac{\overline{\nu}_{i-\frac{1}{2},j+\frac{1}{2}} - \overline{\nu}_{i-\frac{1}{2},j-\frac{1}{2}}}{\Delta\xi\Delta\eta} \right] \right\} \quad (21\text{-}258)$$

and

$$M_\eta^3 = -\left(\frac{1}{Re_\infty}\right) \frac{c_{b2}}{\sigma} (\nu + \overline{\nu}) \left(\eta_x \frac{\partial A}{\partial \eta} + \eta_y \frac{\partial B}{\partial \eta} \right)$$

where

$$- (\nu + \overline{\nu}) \left(\eta_x \frac{\partial A}{\partial \eta} + \eta_y \frac{\partial B}{\partial \eta} \right) \cong$$

$$- (\nu + \overline{\nu})_{i,j} \left\{ (\eta_{x_{i,j}} \eta_{x_{i,j+\frac{1}{2}}} + \eta_{y_{i,j}} \eta_{y_{i,j+\frac{1}{2}}}) \left[\frac{\overline{\nu}_{i,j+1} - \overline{\nu}_{i,j}}{(\Delta\eta)^2} \right] \right\}$$

$$+ (\nu + \overline{\nu})_{i,j} \left\{ (\eta_{x_{i,j}} \eta_{x_{i,j-\frac{1}{2}}} + \eta_{y_{i,j}} \eta_{y_{i,j-\frac{1}{2}}}) \left[\frac{\overline{\nu}_{i,j} - \overline{\nu}_{i,j-1}}{(\Delta\eta)^2} \right] \right\}$$

$$-(\nu+\overline{\nu})_{i,j}\left\{(\eta_{x_{i,j}}\xi_{x_{i,j+\frac{1}{2}}}+\eta_{y_{i,j}}\xi_{y_{i,j+\frac{1}{2}}})\left[\frac{\overline{\nu}_{i+\frac{1}{2},j+\frac{1}{2}}-\overline{\nu}_{i-\frac{1}{2},j+\frac{1}{2}}}{\Delta\xi\Delta\eta}\right]\right\}$$

$$+(\nu+\overline{\nu})_{i,j}\left\{(\eta_{x_{i,j}}\xi_{x_{i,j-\frac{1}{2}}}+\eta_{y_{i,j}}\xi_{y_{i,j-\frac{1}{2}}})\left[\frac{\overline{\nu}_{i+\frac{1}{2},j-\frac{1}{2}}-\overline{\nu}_{i-\frac{1}{2},j-\frac{1}{2}}}{\Delta\xi\Delta\eta}\right]\right\} \quad (21\text{-}259)$$

The production term in this model is simulated by

$$P = c_{b_1}\overline{S}\overline{\nu}(1-f_{t2})$$

where $c_{b_1} = 0.1355$ is a constant, and

$$\overline{S} = S + \frac{\overline{\nu}}{Re_\infty k^2 d^2} f_{v2} \quad , \quad \kappa = 0.41$$

$$f_{v2} = 1 - \frac{\chi}{1+\chi f_{v1}} \quad , \quad \chi = \frac{\overline{\nu}}{\nu}$$

$$f_{v1} = \frac{\chi^3}{\chi^3 + c_{v1}^3} \quad , \quad c_{v1} = 7.1$$

$$f_{t2} = c_{t3}\exp(-c_{t4}\chi^2) \quad , \quad c_{t4} = 2.0$$

At location (i,j),

$$P_{i,j} = c_{b1}\overline{S}_{i,j}\overline{\nu}_{i,j}(1-f_{t2_{i,j}}) \quad (21\text{-}260)$$

$$\overline{S}_{i,j} = S_{i,j} + \frac{\overline{\nu}_{i,j}}{Re_\infty k^2 d_{i,j}^2} f_{v2_{i,j}} \quad (21\text{-}261)$$

$$S_{i,j} = \left|\frac{\partial v}{\partial x} - \frac{\partial u}{\partial y}\right|_{i,j}$$

$$= \left|\xi_x\frac{\partial v}{\partial \xi} + \eta_x\frac{\partial v}{\partial \eta} - \xi_y\frac{\partial u}{\partial \xi} - \eta_y\frac{\partial u}{\partial \eta}\right|_{i,j}$$

$$= \left|\xi_{x_{i,j}}\frac{v_{i+1,j}-v_{i-1,j}}{2\Delta\xi} + \eta_{x_{i,j}}\frac{v_{i,j+1}-v_{i,j-1}}{2\Delta\eta}\right.$$

$$\left. - \xi_{y_{i,j}}\frac{u_{i+1,j}-u_{i-1,j}}{2\Delta\xi} - \eta_{y_{i,j}}\frac{u_{i,j+1}-u_{i,j-1}}{2\Delta\eta}\right| \quad (21\text{-}262)$$

$$\chi_{i,j} = \frac{\overline{\nu}_{i,j}}{\nu_{i,j}} \tag{21-263}$$

$$f_{t2_{i,j}} = c_{t_3} \exp(-c_{t_4} \chi_{i,j}^2) \tag{21-264}$$

$$f_{v2_{i,j}} = 1 - \frac{\chi_{i,j}}{1 + f_{v1_{i,j}}} \tag{21-265}$$

$$f_{v1_{i,j}} = \frac{\chi_{i,j}^3}{\chi_{i,j}^3 + c_{v1}^3} \tag{21-266}$$

The destruction term is given by

$$D = -\frac{1}{Re_\infty}\left(c_{w_1} f_w - \frac{c_{b1}}{k^2} f_{t2}\right)\left[\frac{\overline{\nu}}{d}\right]^2 \tag{21-267}$$

where

$$f_w = g\left(\frac{1 + c_{w_3}^6}{g^6 + c_{w_3}^6}\right)^{\frac{1}{6}} \tag{21-268}$$

$$g = r + c_{w_2}(r^6 - r) \tag{21-269}$$

$$r = \frac{1}{Re_\infty} \frac{\overline{\nu}}{\overline{S}\kappa^2 d^2} \tag{21-270}$$

$$\overline{S} = S + \frac{\overline{\nu}}{Re_\infty \kappa^2 d^2} f_{v_2} \tag{21-271}$$

Thus,

$$D_{i,j} = -\frac{1}{Re_\infty}\left(c_{w_1} f_{w_{i,j}} - \frac{c_{b_1}}{\kappa^2} f_{t2_{i,j}}\right)\left(\frac{\overline{\nu}_{i,j}}{d_{i,j}}\right)^2 \tag{21-272}$$

$$f_{w_{i,j}} = g_{i,j}\left(\frac{1 + c_{w_3}^6}{g_{i,j}^6 + c_{w_3}^6}\right)^{\frac{1}{6}} \tag{21-273}$$

$$g_{i,j} = r_{i,j} + c_{w_2}(r_{i,j}^6 - r_{i,j}) \tag{21-274}$$

$$r_{i,j} = \frac{1}{Re_\infty} \frac{\overline{\nu}_{i,j}}{\overline{S}_{i,j}\kappa^2 d_{i,j}^2} \tag{21-275}$$

The trip term is given by

$$T = Re_\infty f_{t1} (\Delta q)^2$$

where

$$f_{t1} = c_{t1} g_t \exp\left[-c_{t2}\frac{\omega_t^2}{(\Delta q)^2}\left(d^2 + g_t^2 d_t^2\right)\right] \tag{21-276}$$

and

$$g_t = \min\left(0.1\,, \frac{\Delta q}{\omega_t(\Delta x)_t}\right) \tag{21-277}$$

Δq: Is the difference between the velocities at the field point and at the trip point at the wall. Since the velocity at the wall is zero (no suction), Δq is simply the velocity at the field point.

ω_t: Is the wall vorticity at the trip.

Δx_t: Is the grid spacing along the wall at the trip.

d_t: Is the distance from the field point to the trip point.

d: Is the nearest distance from the field point to the wall.

At this point, all the finite difference expressions are in place to develop a numerical scheme. The ADI formulation, which was introduced in Chapter 3, will be used for this purpose. Recall that the governing equation given by (21-139) is

$$\frac{\partial \overline{\nu}}{\partial t} = M + P - D + T \tag{21-278}$$

Furthermore, Equation (21-278) is nonlinear due to the terms in M, P, and D. Thus, before a numerical scheme is considered, Equation (21-278) must be linearized. The linearization procedure developed in Chapter 6, and similarly developed in Chapter 11, and given by Relation (11-81), is used to linearize Equation (21-278). The general expression is

$$E^{n+1} = E^n + A\Delta u \tag{21-279}$$

where

$$A = \frac{\partial E}{\partial u}$$

Now, to develop an implicit scheme, Equation (21-278) is applied at time level $n+1$. Therefore,

$$\left(\frac{\partial \overline{\nu}}{\partial t}\right)^{n+1} = M^{n+1} + P^{n+1} - D^{n+1} + T^n \tag{21-280}$$

where the trip term is treated as a source term evaluated at time level n. Now the linearization (21-279) provides

$$M^{n+1} = M^n + \frac{\partial M}{\partial \overline{\nu}}\Delta\overline{\nu} = M^n + \overline{M}\Delta\overline{\nu} \tag{21-281}$$

$$P^{n+1} = P^n + \frac{\partial P}{\partial \overline{\nu}}\Delta\overline{\nu} = P^n + \overline{P}\Delta\overline{\nu} \tag{21-282}$$

$$D^{n+1} = D^n + \frac{\partial D}{D\overline{\nu}}\Delta\overline{\nu} = D^n + \overline{D}\Delta\overline{\nu} \tag{21-283}$$

Now Equation (21-280) can be written as

$$\frac{\Delta\overline{\nu}}{\Delta t} - (\overline{M} + \overline{P} - \overline{D})\Delta\overline{\nu} = M^n + P^n - D^n + T^n \tag{21-284}$$

or

$$\left[1 - (\overline{M} + \overline{P} - \overline{D})\Delta t\right]\Delta\overline{\nu} = (M^n + P^n - D^n + T^n)\Delta t \tag{21-285}$$

Recall that, in the development of finite difference expressions for M, it was decomposed as M_ξ and M_η. Therefore, the Equation (21-285) can be written as

$$\left[1 - (\overline{M}_\xi + \overline{M}_\eta + \overline{P} - \overline{D})\Delta t\right]\Delta\overline{\nu} = (M^n + P^n - D^n + T^n)\Delta t$$

Now this equation is factored as

$$\left[1 - \Delta t\overline{M}_\xi\right]\left[1 - \Delta t\left(\overline{M}_\eta + \overline{P} - \overline{D}\right)\right]\Delta\overline{\nu} =$$

$$\left(M_\xi^n + M_\eta^n + P^n - D^n + T^n\right)\Delta t$$

and solved sequentially with the following equations

$$(1 - \Delta t\,\overline{M}_\xi)\,\Delta\overline{\nu}^* = RHS \tag{21-286}$$

$$\left[1 - \Delta t\,(\overline{M}_\eta + \overline{P} - \overline{D})\right]\Delta\overline{\nu} = \Delta\overline{\nu}^* \tag{21-287}$$

As seen previously, each one of these equations, once applied to all of the grid points, will result in a tridiagonal system. For example, Equation (21-286) will yield

$$(1 - \Delta t\,\overline{M}_\xi)_{i-1,j}\Delta\overline{\nu}^*_{i-1,j} + (1 - \Delta t\,\overline{M}_\xi)_{i,j}\Delta\overline{\nu}^*_{i,j}$$

$$+ (1 - \Delta t\,\overline{M}_\xi)_{i+1,j}\Delta\overline{\nu}^*_{i+1,j} = RHS_{i,j} \tag{21-288}$$

The finite difference expressions for the terms in $RHS_{i,j}$ of Equation (21-288) have already been identified. The "bar" quantities are determined in a similar fashion as that of Section 21.6.1.1. For example, the results for $\overline{M}_\xi^1$ are as follows

$$\left.\overline{M}_\xi^1\right|_{i+1,j} = -\frac{1}{2}(U_{i,j} - |U_{i,j}|)\left(\frac{1}{\Delta\xi}\right)$$

$$\left.\overline{M}_\xi^1\right|_{i,j} = -\left\{|U_{i,j}|\left(\frac{1}{\Delta\xi}\right)\right\}$$

$$\left.\overline{M}_\xi^1\right|_{i-1,j} = \frac{1}{2}(U_{i,j} + |U_{i,j}|)\left(\frac{1}{\Delta\xi}\right)$$

Similar expressions can be obtained for the remaining terms.

Once the tridiagonal system of equations given by (21-286) is solved, the right-hand side of (21-287) is known, and, subsequently, it can be solved. The system of tridiagonal equations given by (21-286) or (21-287) can be solved by any tridiagonal solver such as the scheme discussed in Appendix B.

21.6.2.2 Explicit Formulation: The explicit formulation of Equation (21-127) is similar to that of (21-252) and is given by

$$\begin{aligned}
\overline{\nu}_{i,j}^{n+1} = {} & \overline{\nu}_{i,j}^{n} - \Delta t\left\{\frac{1}{2}(U_{i,j} + |U_{i,j}|)\left(\frac{\overline{\nu}_{i,j} - \overline{\nu}_{i-1,j}}{\Delta\xi}\right) + \frac{1}{2}(U_{i,j} - |U_{i,j}|)\left(\frac{\overline{\nu}_{i+1,j} - \overline{\nu}_{i,j}}{\Delta\xi}\right)\right. \\
& \left. + \frac{1}{2}(V_{i,j} + |V_{i,j}|)\left(\frac{\overline{\nu}_{i,j} - \overline{\nu}_{i,j-1}}{\Delta\eta}\right) + \frac{1}{2}(V_{i,j} - |V_{i,j}|)\left(\frac{\overline{\nu}_{i,j+1} - \overline{\nu}_{i,j}}{\Delta\eta}\right)\right\} \\
& + \Delta t\left(\frac{1}{Re_\infty}\right)\left(\frac{1 + c_{b2}}{\sigma}\right)\left[(\nu + \overline{\nu})_{i+\frac{1}{2},j}\left\{(\xi_{x_{i,j}}\xi_{x_{i+\frac{1}{2},j}} + \xi_{y_{i,j}}\xi_{y_{i+\frac{1}{2},j}})\left[\frac{\overline{\nu}_{i+1,j} - \overline{\nu}_{i,j}}{(\Delta\xi)^2}\right]\right\}\right. \\
& - (\nu + \overline{\nu})_{i-\frac{1}{2},j}\left\{(\xi_{x_{i,j}}\xi_{x_{i-\frac{1}{2},j}} + \xi_{y_{i,j}}\xi_{y_{i-\frac{1}{2},j}})\left[\frac{\overline{\nu}_{i,j} - \overline{\nu}_{i-1,j}}{(\Delta\xi)^2}\right]\right\} \\
& + (\nu + \overline{\nu})_{i+\frac{1}{2},j}\left\{(\xi_{x_{i,j}}\eta_{x_{i+\frac{1}{2},j}} + \xi_{y_{i,j}}\eta_{y_{i+\frac{1}{2},j}})\left[\frac{\overline{\nu}_{i+\frac{1}{2},j+\frac{1}{2}} - \overline{\nu}_{i+\frac{1}{2},j-\frac{1}{2}}}{\Delta\xi\Delta\eta}\right]\right\} \\
& - (\nu + \overline{\nu})_{i-\frac{1}{2},j}\left\{(\xi_{x_{i,j}}\eta_{x_{i-\frac{1}{2},j}} + \xi_{y_{i,j}}\eta_{y_{i-\frac{1}{2},j}})\left[\frac{\overline{\nu}_{i-\frac{1}{2},j+\frac{1}{2}} - \overline{\nu}_{i-\frac{1}{2},j-\frac{1}{2}}}{\Delta\xi\Delta\eta}\right]\right\} \\
& + (\nu + \overline{\nu})_{i,j+\frac{1}{2}}\left\{(\eta_{x_{i,j}}\eta_{x_{i,j+\frac{1}{2}}} + \eta_{y_{i,j}}\eta_{y_{i,j+\frac{1}{2}}})\left[\frac{\overline{\nu}_{i,j+1} - \overline{\nu}_{i,j}}{(\Delta\eta)^2}\right]\right\}
\end{aligned}$$

$$
\begin{aligned}
&- (\nu + \bar{\nu})_{i,j-\frac{1}{2}} \left\{ (\eta_{x_{i,j}} \eta_{x_{i,j-\frac{1}{2}}} + \eta_{y_{i,j}} \eta_{y_{i,j-\frac{1}{2}}}) \left[\frac{\bar{\nu}_{i,j} - \bar{\nu}_{i,j-1}}{(\Delta\eta)^2} \right] \right\} \\
&+ (\nu + \bar{\nu})_{i,j+\frac{1}{2}} \left\{ (\eta_{x_{i,j}} \xi_{x_{i,j+\frac{1}{2}}} + \eta_{y_{i,j}} \xi_{y_{i,j+\frac{1}{2}}}) \left[\frac{\bar{\nu}_{i+\frac{1}{2},j+\frac{1}{2}} - \bar{\nu}_{i-\frac{1}{2},j+\frac{1}{2}}}{\Delta\xi\Delta\eta} \right] \right\} \\
&- (\nu + \bar{\nu})_{i,j-\frac{1}{2}} \left\{ (\eta_{x_{i,j}} \xi_{x_{i,j-\frac{1}{2}}} + \eta_{y_{i,j}} \xi_{y_{i,j-\frac{1}{2}}}) \left[\frac{\bar{\nu}_{i+\frac{1}{2},j-\frac{1}{2}} - \bar{\nu}_{i-\frac{1}{2},j-\frac{1}{2}}}{\Delta\xi\Delta\eta} \right] \right\} \Bigg] \\
&+ \Delta t \left(\frac{1}{Re_\infty} \right) \left(\frac{c_{b2}}{\sigma} \right) \Bigg[-(\nu + \bar{\nu})_{i,j} \left\{ (\xi_{x_{i,j}} \xi_{x_{i+\frac{1}{2},j}} + \xi_{y_{i,j}} \xi_{y_{i+\frac{1}{2},j}}) \left[\frac{\bar{\nu}_{i+1,j} - \bar{\nu}_{i,j}}{(\Delta\xi)^2} \right] \right\} \\
&+ (\nu + \bar{\nu})_{i,j} \left\{ (\xi_{x_{i,j}} \xi_{x_{i-\frac{1}{2},j}} + \xi_{y_{i,j}} \xi_{y_{i-\frac{1}{2},j}}) \left[\frac{\bar{\nu}_{i,j} - \bar{\nu}_{i-1,j}}{(\Delta\xi)^2} \right] \right\} \\
&- (\nu + \bar{\nu})_{i,j} \left\{ (\xi_{x_{i,j}} \eta_{x_{i+\frac{1}{2},j}} + \xi_{y_{i,j}} \eta_{y_{i+\frac{1}{2},j}}) \left[\frac{\bar{\nu}_{i+\frac{1}{2},j+\frac{1}{2}} - \bar{\nu}_{i+\frac{1}{2},j-\frac{1}{2}}}{\Delta\xi\Delta\eta} \right] \right\} \\
&+ (\nu + \bar{\nu})_{i,j} \left\{ (\xi_{x_{i,j}} \eta_{x_{i-\frac{1}{2},j}} + \xi_{y_{i,j}} \eta_{y_{i-\frac{1}{2},j}}) \left[\frac{\bar{\nu}_{i-\frac{1}{2},j+\frac{1}{2}} - \bar{\nu}_{i-\frac{1}{2},j-\frac{1}{2}}}{\Delta\xi\Delta\eta} \right] \right\} \\
&- (\nu + \bar{\nu})_{i,j} \left\{ (\eta_{x_{i,j}} \eta_{x_{i,j+\frac{1}{2}}} + \eta_{y_{i,j}} \eta_{y_{i,j+\frac{1}{2}}}) \left[\frac{\bar{\nu}_{i,j+1} - \bar{\nu}_{i,j}}{(\Delta\eta)^2} \right] \right\} \\
&+ (\nu + \bar{\nu})_{i,j} \left\{ (\eta_{x_{i,j}} \eta_{x_{i,j-\frac{1}{2}}} + \eta_{y_{i,j}} \eta_{y_{i,j-\frac{1}{2}}}) \left[\frac{\bar{\nu}_{i,j} - \bar{\nu}_{i,j-1}}{(\Delta\eta)^2} \right] \right\} \\
&- (\nu + \bar{\nu})_{i,j} \left\{ (\eta_{x_{i,j}} \xi_{x_{i,j+\frac{1}{2}}} + \eta_{y_{i,j}} \xi_{y_{i,j+\frac{1}{2}}}) \left[\frac{\bar{\nu}_{i+\frac{1}{2},j+\frac{1}{2}} - \bar{\nu}_{i-\frac{1}{2},j+\frac{1}{2}}}{\Delta\xi\Delta\eta} \right] \right\} \\
&+ (\nu + \bar{\nu})_{i,j} \left\{ (\eta_{x_{i,j}} \xi_{x_{i,j-\frac{1}{2}}} + \eta_{y_{i,j}} \xi_{y_{i,j-\frac{1}{2}}}) \left[\frac{\bar{\nu}_{i+\frac{1}{2},j-\frac{1}{2}} - \bar{\nu}_{i-\frac{1}{2},j-\frac{1}{2}}}{\Delta\xi\Delta\eta} \right] \right\} \Bigg] \\
&+ \Delta t \Bigg\{ c_{b_1} \overline{S}_{i,j} \bar{\nu}_{i,j} (1 - f_{t2_{i,j}}) \\
&- \frac{1}{Re_\infty} \left(c_{w1} f_{w_{i,j}} - \frac{c_{b_1}}{\kappa^2} f_{t2_{i,j}} \right) \left(\frac{\bar{\nu}_{i,j}}{d_{i,j}} \right)^2 + Re_\infty f_{t1} (\Delta q_{i,j})^2 \Bigg\} \qquad \text{(21-289)}
\end{aligned}
$$

21.6.3 Two-Equation Turbulence Models

Similar to the one-equation models, the convective terms are approximated by either a first-order or a second-order upwind scheme. Second-order central difference approximation is used for the remaining terms either at the grid points or at midpoints. If central difference approximation is applied at the points next to the boundaries, a difficulty will be encountered due to the appearance of points in the finite difference equation which are outside the domain. At these points, one-sided approximation can be used to eliminate the difficulty. When approximation at the midpoints is used, this difficulty does not occur. Both formulations are provided in this section. In the following formulations, the first-order upwind scheme is used. Furthermore, since the many terms in the k-equation, ϵ-equation, and ω-equation are similar, only the finite difference equation for the k-equation is provided in detail. Subsequently, the extra terms in the ϵ-equation and the ω-equation are investigated. The k-equation given by (21-205) is divided by ρ and written as

$$\begin{aligned}\frac{\partial k}{\partial t} &= -U\frac{\partial k}{\partial \xi} - V\frac{\partial k}{\partial \eta} + \frac{1}{\rho\, Re_\infty}\left\{\xi_x \frac{\partial}{\partial \xi}\left[(MUS)\,(KX)\right]\right. \\ &\quad + \eta_x \frac{\partial}{\partial \eta}\left[(MUS)\,(KX)\right] + \xi_y \frac{\partial}{\partial \xi}\left[(MUS)\,(KY)\right] \\ &\quad \left. + \eta_y \frac{\partial}{\partial \eta}\left[(MUS)\,(KY)\right]\right\} + \frac{P_k}{\rho} - \beta^* \omega\, k \qquad (21\text{-}290)\end{aligned}$$

Now an explicit formulation for Equation (21-290) is written as follows

$$\begin{aligned}k_{i,j}^{n+1} &= k_{i,j}^n - \Delta t\left[\left\{\frac{1}{2}(U_{i,j} + |U_{i,j}|)\,\frac{k_{i,j}^n - k_{i-1,j}^n}{\Delta \xi} + \right.\right. \\ &\quad + \frac{1}{2}\left(U_{i,j} - |U_{i,j}|\right)\frac{k_{i+1,j}^n - k_{i,j}^n}{\Delta \xi} \\ &\quad + \frac{1}{2}\left(V_{i,j} + |V_{i,j}|\right)\frac{k_{i,j}^n - k_{i,j-1}^n}{\Delta \eta} \\ &\quad + \frac{1}{2}\left(V_{i,j} - |V_{i,j}|\right)\frac{k_{i,j+1}^n - k_{i,j}^n}{\Delta \eta} \\ &\quad + \xi_{x_{i,j}}\frac{1}{\rho_{i,j} Re_\infty}\left(\frac{1}{2\Delta\xi}\right)\left\{\left[(MUS)\,(KX)\right]_{i+1,j} - \left[(MUS)\,(KX)\right]_{i-1,j}\right\}\end{aligned}$$

$$+ \eta_{x_{i,j}} \frac{1}{\rho_{i,j} Re_\infty} \left(\frac{1}{2\Delta\eta} \right) \left\{ [(MUS)(KX)]_{i,j+1} - [(MUS)(KX)]_{i,j-1} \right\}$$

$$+ \xi_{y_{i,j}} \frac{1}{\rho_{i,j} Re_\infty} \left(\frac{1}{2\Delta\xi} \right) \left\{ [(MUS)(KY)]_{i+1,j} - [(MUS)(KY)]_{i-1,j} \right\}$$

$$+ \eta_{y_{i,j}} \frac{1}{\rho_{i,j} Re_\infty} \left(\frac{1}{2\Delta\eta} \right) \left\{ [(MUS)(KY)]_{i,j+1} - [(MUS)(KY)]_{i,j-1} \right\}$$

$$\left. + \frac{1}{\rho_{i,j}} (P_k)_{i,j} - \beta^* \omega_{i,j} k_{i,j} \right] \tag{21-291}$$

where the production term P_k, as given by (21-220) is computed as follows

$$\begin{aligned} (P_k)_{i,j} &= (P1)_{i,j} \frac{u_{i+1,j} - u_{i-1,j}}{2\Delta\xi} + (P2)_{i,j} \frac{u_{i,j+1} - u_{i,j-1}}{2\Delta\eta} \\ &\quad + (P3)_{i,j} \frac{v_{i+1,j} - v_{i-1,j}}{2\Delta\xi} + (P4)_{i,j} \frac{v_{i,j+1} - v_{i,j-1}}{2\Delta\eta} \end{aligned}$$

To illustrate the midpoint approximation, consider the first four terms on the right-hand side of Equation (21-205). Keep in mind that the convective terms are approximated by upwind scheme, as before. The four terms of (21-205) under consideration are

$$\frac{1}{Re_\infty} \xi_x \frac{\partial}{\partial \xi} \left[(MUS) \left(\xi_x \frac{\partial k}{\partial \xi} + \eta_x \frac{\partial k}{\partial \eta} \right) \right]$$

$$+ \frac{1}{Re_\infty} \eta_x \frac{\partial}{\partial \eta} \left[(MUS) \left(\xi_x \frac{\partial k}{\partial \xi} + \eta_x \frac{\partial k}{\partial \eta} \right) \right]$$

$$+ \frac{1}{Re_\infty} \xi_y \frac{\partial}{\partial \xi} \left[(MUS) \left(\xi_y \frac{\partial k}{\partial \xi} + \eta_y \frac{\partial k}{\partial \eta} \right) \right]$$

$$+ \frac{1}{Re_\infty} \eta_y \frac{\partial}{\partial \eta} \left[(MUS) \left(\xi_y \frac{\partial k}{\partial \xi} + \eta_y \frac{\partial k}{\partial \eta} \right) \right]$$

$$= \frac{1}{Re_\infty} \left\{ \xi_x \frac{\partial A}{\partial \xi} + \xi_y \frac{\partial B}{\partial \xi} \right\} + \frac{1}{Re_\infty} \left\{ \eta_x \frac{\partial A}{\partial \eta} + \eta_y \frac{\partial B}{\partial \eta} \right\} \tag{21-292}$$

where

$$A = (MUS)\left(\xi_x \frac{\partial k}{\partial \xi} + \eta_x \frac{\partial k}{\partial \eta}\right)$$

and

$$B = (MUS)\left(\xi_y \frac{\partial k}{\partial \xi} + \eta_y \frac{\partial k}{\partial \eta}\right)$$

Now consider the finite difference expression for $(\xi_x \partial A/\partial \xi)$ where central difference approximation is used.

$$\left(\xi_x \frac{\partial A}{\partial \xi}\right)_{i,j} = \xi_{x_{i,j}} \frac{A_{i+\frac{1}{2},j} - A_{i-\frac{1}{2},j}}{\Delta \xi} \tag{21-293}$$

where

$$\begin{aligned} A_{i+\frac{1}{2},j} &= (MUS)_{i+\frac{1}{2},j}\left(\xi_x \frac{\partial k}{\partial \xi} + \eta_x \frac{\partial k}{\partial \eta}\right)_{i+\frac{1}{2},j} \\ &= (MUS)_{i+\frac{1}{2},j}\left\{\xi_{x_{i+\frac{1}{2},j}} \frac{k_{i+1,j} - k_{i,j}}{\Delta \xi} + \eta_{x_{i+\frac{1}{2},j}} \frac{k_{i+\frac{1}{2},j+\frac{1}{2}} - k_{i+\frac{1}{2},j-\frac{1}{2}}}{\Delta \eta}\right\} \end{aligned} \tag{21-294}$$

and

$$\begin{aligned} A_{i-\frac{1}{2},j} &= (MUS)_{i-\frac{1}{2},j}\left\{\xi_{x_{i-\frac{1}{2},j}} \frac{k_{i,j} - k_{i-1,j}}{\Delta \xi}\right. \\ &\qquad \left. + \eta_{x_{i-\frac{1}{2},j}} \frac{k_{i-\frac{1}{2},j+\frac{1}{2}} - k_{i-\frac{1}{2},j-\frac{1}{2}}}{\Delta \eta}\right\} \end{aligned} \tag{21-295}$$

Similarly,

$$B_{i+\frac{1}{2},j} = (MUS)_{i+\frac{1}{2},j}\left\{\xi_{y_{i+\frac{1}{2},j}} \frac{k_{i+1,j} - k_{i,j}}{\Delta \xi} + \eta_{y_{i+\frac{1}{2},j}} \frac{k_{i+\frac{1}{2},j+\frac{1}{2}} - k_{i+\frac{1}{2},j-\frac{1}{2}}}{\Delta \eta}\right\} \tag{21-296}$$

$$B_{i-\frac{1}{2},j} = (MUS)_{i-\frac{1}{2},j}\left\{\xi_{y_{i-\frac{1}{2},j}} \frac{k_{i,j} - k_{i-1,j}}{\Delta \xi} + \eta_{y_{i-\frac{1}{2},j}} \frac{k_{i-\frac{1}{2},j+\frac{1}{2}} - k_{i-\frac{1}{2},j-\frac{1}{2}}}{\Delta \eta}\right\} \tag{21-297}$$

Now the first term of expression (21-292) becomes

$$\frac{1}{Re_\infty}\left\{\xi_x \frac{\partial A}{\partial \xi} + \xi_y \frac{\partial B}{\partial \xi}\right\} =$$

$$\frac{1}{Re_\infty}\left\{(MUS)_{i+\frac{1}{2},j} \frac{\xi_{x_{i,j}} \xi_{x_{i+\frac{1}{2},j}}}{(\Delta \xi)^2}(k_{i+1,j} - k_{i,j})\right.$$

$$+ (MUS)_{i+\frac{1}{2},j} \frac{\xi_{x_{i,j}} \eta_{x_{i+\frac{1}{2},j}}}{\Delta\xi\Delta\eta} (k_{i+\frac{1}{2},j+\frac{1}{2}} - k_{i+\frac{1}{2},j-\frac{1}{2}})$$

$$- (MUS)_{i-\frac{1}{2},j} \frac{\xi_{x_{i,j}} \xi_{x_{i-\frac{1}{2},j}}}{(\Delta\xi)^2} (k_{i,j} - k_{i-1,j})$$

$$- (MUS)_{i-\frac{1}{2},j} \frac{\xi_{x_{i,j}} \eta_{x_{i-\frac{1}{2},j}}}{\Delta\xi\Delta\eta} (k_{i-\frac{1}{2},j+\frac{1}{2}} - k_{i-\frac{1}{2},j-\frac{1}{2}})$$

$$+ (MUS)_{i+\frac{1}{2},j} \frac{\xi_{y_{i,j}} \xi_{y_{i+\frac{1}{2},j}}}{(\Delta\xi)^2} (k_{i+1,j} - k_{i,j})$$

$$+ (MUS)_{i+\frac{1}{2},j} \frac{\xi_{y_{i,j}} \eta_{y_{i+\frac{1}{2},j}}}{\Delta\xi\Delta\eta} (k_{i+\frac{1}{2},j+\frac{1}{2}} - k_{i+\frac{1}{2},j-\frac{1}{2}})$$

$$- (MUS)_{i-\frac{1}{2},j} \frac{\xi_{y_{i,j}} \xi_{y_{i-\frac{1}{2},j}}}{(\Delta\xi)^2} (k_{i,j} - k_{i-1,j})$$

$$- (MUS)_{i-\frac{1}{2},j} \frac{\xi_{y_{i,j}} \eta_{y_{i-\frac{1}{2},j}}}{\Delta\xi\Delta\eta} (k_{i-\frac{1}{2},j+\frac{1}{2}} - k_{i-\frac{1}{2},j-\frac{1}{2}}) \Bigg\}$$

$$= \frac{1}{Re_\infty} \Bigg\{ (MUS)_{i+\frac{1}{2},j} \frac{1}{(\Delta\xi)^2} \left(\xi_{x_{i,j}} \xi_{x_{i+\frac{1}{2},j}} + \xi_{y_{i,j}} \xi_{y_{i+\frac{1}{2},j}} \right) (k_{i+1,j} - k_{i,j})$$

$$- (MUS)_{i-\frac{1}{2},j} \frac{1}{(\Delta\xi)^2} \left(\xi_{x_{i,j}} \xi_{x_{i-\frac{1}{2},j}} + \xi_{y_{i,j}} \xi_{y_{i-\frac{1}{2},j}} \right) (k_{i,j} - k_{i-1,j})$$

$$+ (MUS)_{i+\frac{1}{2},j} \frac{1}{\Delta\xi\Delta\eta} \left(\xi_{x_{i,j}} \eta_{x_{i+\frac{1}{2},j}} + \xi_{y_{i,j}} \eta_{y_{i+\frac{1}{2},j}} \right) \left(k_{i+\frac{1}{2},j+\frac{1}{2}} - k_{i+\frac{1}{2},j-\frac{1}{2}} \right)$$

$$- (MUS)_{i-\frac{1}{2},j} \frac{1}{\Delta\xi\Delta\eta} \left(\xi_{x_{i,j}} \eta_{x_{i-\frac{1}{2},j}} + \xi_{y_{i,j}} \eta_{y_{i-\frac{1}{2},j}} \right) \left(k_{i-\frac{1}{2},j+\frac{1}{2}} - k_{i-\frac{1}{2},j-\frac{1}{2}} \right) \Bigg\} \quad (21\text{-}298)$$

The midpoint values are determined according to

$$(\ )_{i\pm\frac{1}{2},j} = \frac{1}{2} [(\ )_{i,j} + (\ )_{i\pm1,j}]$$

$$(\ )_{i\pm\frac{1}{2},j+\frac{1}{2}} = \frac{1}{4} [(\ )_{i,j} + (\ )_{i\pm1,j} + (\ )_{i\pm1,j+1} + (\ )_{i,j+1}]$$

$$(\ )_{i\pm\frac{1}{2},j-\frac{1}{2}} = \frac{1}{4}[(\ )_{i,j} + (\ )_{i\pm1,j} + (\ )_{i,j-1} + (\ )_{i\pm1,j-1}]$$

For example,

$$k_{i+\frac{1}{2},j+\frac{1}{2}} = \frac{1}{4}(k_{i,j} + k_{i+1,j} + k_{i+1,j+1} + k_{i,j+1})$$

Similarly, the second term of (21-292) which is given by

$$\frac{1}{Re_\infty}\left\{\eta_x\frac{\partial A}{\partial\eta} + \eta_y\frac{\partial B}{\partial\eta}\right\}$$

is approximated by

$$\frac{1}{Re_\infty}\left\{(MUS)_{i,j+\frac{1}{2}}\frac{1}{(\Delta\eta)^2}\left(\eta_{x_{i,j}}\eta_{x_{i,j+\frac{1}{2}}} + \eta_{y_{i,j}}\eta_{y_{i,j+\frac{1}{2}}}\right)\left(k_{i,j+\frac{1}{2}} - k_{i,j}\right)\right.$$

$$- (MUS)_{i,j-\frac{1}{2}}\frac{1}{(\Delta\eta)^2}\left(\eta_{x_{i,j}}\eta_{x_{i,j-\frac{1}{2}}} + \eta_{y_{i,j}}\eta_{y_{i,j-\frac{1}{2}}}\right)(k_{i,j} - k_{i,j-1})$$

$$+ (MUS)_{i,j+\frac{1}{2}}\frac{1}{\Delta\xi\Delta\eta}\left(\eta_{x_{i,j}}\xi_{x_{i,j+\frac{1}{2}}} + \eta_{y_{i,j}}\xi_{y_{i,j+\frac{1}{2}}}\right)\left(k_{i+\frac{1}{2},j+\frac{1}{2}} - k_{i-\frac{1}{2},j+\frac{1}{2}}\right)$$

$$\left.- (MUS)_{i,j-\frac{1}{2}}\frac{1}{\Delta\xi\Delta\eta}\left(\eta_{x_{i,j}}\xi_{x_{i,j-\frac{1}{2}}} + \eta_{y_{i,j}}\xi_{y_{i,j-\frac{1}{2}}}\right)\left(k_{i+\frac{1}{2},j-\frac{1}{2}} - k_{i-\frac{1}{2},j-\frac{1}{2}}\right)\right\} \quad (21\text{-}299)$$

Finally, the finite difference equation for the k-equation becomes,

$$k_{i,j}^{n+1} = k_{i,j}^n - \Delta t\left[\left\{\frac{1}{2}(U_{i,j} + |U_{i,j}|)\frac{k_{i,j}^n - k_{i-1,j}^n}{\Delta\xi}\right.\right.$$

$$+ \frac{1}{2}(U_{i,j} - |U_{i,j}|)\frac{k_{i+1,j}^n - k_{i,j}^n}{\Delta\xi} + \frac{1}{2}(V_{i,j} + |V_{i,j}|)\frac{k_{i,j}^n - k_{i,j-1}^n}{\Delta\eta}$$

$$\left. + \frac{1}{2}(V_{i,j} - |V_{i,j}|)\frac{k_{i,j+1}^n - k_{i,j}^n}{\Delta\eta}\right\}$$

$$+ \frac{1}{\rho_{i,j}Re_\infty}\left\{(MUS)_{i+\frac{1}{2},j}\frac{1}{(\Delta\xi)^2}\left(\xi_{x_{i,j}}\xi_{x_{i+\frac{1}{2},j}} + \xi_{y_{i,j}}\xi_{y_{i+\frac{1}{2},j}}\right)(k_{i+1,j} - k_{i,j})\right.$$

$$- (MUS)_{i-\frac{1}{2},j}\frac{1}{(\Delta\xi)^2}\left(\xi_{x_{i,j}}\xi_{x_{i-\frac{1}{2},j}} + \xi_{y_{i,j}}\xi_{y_{i-\frac{1}{2},j}}\right)(k_{i,j} - k_{i-1,j})$$

$$+ (MUS)_{i+\frac{1}{2},j}\frac{1}{\Delta\xi\Delta\eta}\left(\xi_{x_{i,j}}\eta_{x_{i+\frac{1}{2},j}} + \xi_{y_{i,j}}\eta_{y_{i+\frac{1}{2},j}}\right)$$

$$\left[(0.25)\left(k_{i,j+1}+k_{i+1,j+1}-k_{i,j-1}-k_{i+1,j-1}\right)\right]$$

$$-(MUS)_{i-\frac{1}{2}j}\frac{1}{\Delta\xi\Delta\eta}\left(\xi_{x_{i,j}}\eta_{x_{i-\frac{1}{2}j}}+\xi_{y_{i,j}}\eta_{y_{i-\frac{1}{2}j}}\right)$$

$$\left[(0.25)(k_{i,j+1}+k_{i-1,j+1}-k_{i,j-1}-k_{i-1,j-1})\right]$$

$$+(MUS)_{i,j+\frac{1}{2}}\frac{1}{(\Delta\eta)^2}\left(\eta_{x_{i,j}}\eta_{x_{i,j+\frac{1}{2}}}+\eta_{y_{i,j}}\eta_{y_{i,j+\frac{1}{2}}}\right)(k_{i,j+1}-k_{i,j})$$

$$-(MUS)_{i,j-\frac{1}{2}}\frac{1}{(\Delta\eta)^2}\left(\eta_{x_{i,j}}\eta_{x_{i,j-\frac{1}{2}}}+\eta_{y_{i,j}}\eta_{y_{i,j-\frac{1}{2}}}\right)(k_{i,j}-k_{i,j-1})$$

$$+(MUS)_{i,j+\frac{1}{2}}\frac{1}{\Delta\xi\Delta\eta}\left(\eta_{x_{i,j}}\xi_{x_{i,j+\frac{1}{2}}}+\eta_{y_{i,j}}\xi_{y_{i,j+\frac{1}{2}}}\right)$$

$$\left[(0.25)\left(k_{i+1,j}+k_{i+1,j+1}-k_{i-1,j}-k_{i-1,j+1}\right)\right]-$$

$$-(MUS)_{i,j-\frac{1}{2}}\frac{1}{\Delta\xi\Delta\eta}\left(\eta_{x_{i,j}}\xi_{x_{i,j-\frac{1}{2}}}+\eta_{y_{i,j}}\xi_{y_{i,j-\frac{1}{2}}}\right)$$

$$\left.\left[(0.25)\left(k_{i+1,j}+k_{i+1,j-1}-k_{i-1,j}-k_{i-1,j-1}\right)\right]\right\}$$

$$\left.+\left\{\frac{1}{\rho_{i,j}Re_\infty}(P_k)_{i,j}-\beta^*\omega_{i,j}k_{i,j}\right\}\right] \tag{21-300}$$

The ω-equation has similar terms as the k-equation and, therefore, the FDE just investigated can be used for the ω-equation by replacing k with ω. However, recall that the ω-equation has an extra term which will be investigated at this point.

The seventh term of the ω-equation given by (21-209) is

$$\left[2\rho(1-F_1)\sigma_{\omega 2}\frac{1}{\omega}\right]\left[\left(\xi_x\frac{\partial k}{\partial\xi}+\eta_x\frac{\partial k}{\partial\eta}\right)\left(\xi_x\frac{\partial\omega}{\partial\xi}+\eta_x\frac{\partial\omega}{\partial\eta}\right)\right.$$

$$\left.+\left(\xi_y\frac{\partial k}{\partial\xi}+\eta_y\frac{\partial k}{\partial\eta}\right)\left(\xi_y\frac{\partial\omega}{\partial\xi}+\eta_y\frac{\partial\omega}{\partial\eta}\right)\right]$$

The second bracket can be rearranged as

$$(\xi_x^2+\xi_y^2)\frac{\partial k}{\partial\xi}\frac{\partial\omega}{\partial\xi}+(\xi_x\eta_x+\xi_y\eta_y)\frac{\partial k}{\partial\xi}\frac{\partial\omega}{\partial\eta}$$

$$+ (\xi_x \eta_x + \xi_y \eta_y) \frac{\partial k}{\partial \eta} \frac{\partial \omega}{\partial \xi} + (\eta_x^2 + \eta_y^2) \frac{\partial k}{\partial \eta} \frac{\partial \omega}{\partial \eta}$$

$$= a_4 \frac{\partial k}{\partial \xi} \frac{\partial \omega}{\partial \xi} + c_5 \left(\frac{\partial k}{\partial \xi} \frac{\partial \omega}{\partial \eta} + \frac{\partial k}{\partial \eta} \frac{\partial \omega}{\partial \xi} \right) + b_4 \frac{\partial k}{\partial \eta} \frac{\partial \omega}{\partial \eta}$$

The finite difference for the expression above is written as

$$a_4 \frac{(k_{i+1,j} k_{i-1,j}) (\omega_{i+1,j} - \omega_{i-1,j})}{4(\Delta\xi)^2}$$

$$+ c_{5_{i,j}} \frac{(k_{i+1,j} - k_{i-1,j}) (\omega_{i,j+1} - \omega_{i,j-1}) + (k_{i,j+1} - k_{i,j-1}) (\omega_{i+1,j} - \omega_{i-1,j})}{4(\Delta\xi) (\Delta\eta)}$$

$$+ b_{4_{i,j}} \frac{(k_{i,j+1} - k_{i,j-1}) (\omega_{i,j+1} - \omega_{i,j-1})}{4(\Delta\eta)^2} \tag{21-301}$$

21.7 Applications

Several turbulence models have been introduced in this chapter. The applications of these models to three different flowfields are illustrated in this section. Computations are performed with several turbulence models for each domain such that the solutions can be compared to each other and the experimental data.

The flowfield is solved either by the first-order and/or second-order flux-vector splitting schemes described in Section 14.4.1.2 and given by Equations (14-21) and (14-22) or by the modified fourth-order Runge-Kutta scheme described in Section 14.4.1.3.

21.7.1 Shock/Boundary Layer Interaction

The oblique shock generated at a compression corner and its interaction with turbulent boundary layer developed near the surface occurs in several engineering applications. Thus, the prediction of such a complex flowfield is essential for design and analysis purposes.

Consider now a two-dimensional compression corner in a supersonic flow. Several physical observations have been made with regard to this type of flowfield as illustrated in Figure (21-7). (1) At the corner a shock will form causing rise in pressure. From basic compressible fluid mechanics it is well known that the strength of the shock and subsequent rise in pressure will increase as the angle of compression corner, and/or the free stream Mach number is increased. (2) A consequence of the flow separation is the appearance of a stagnation point at the reattachment location.

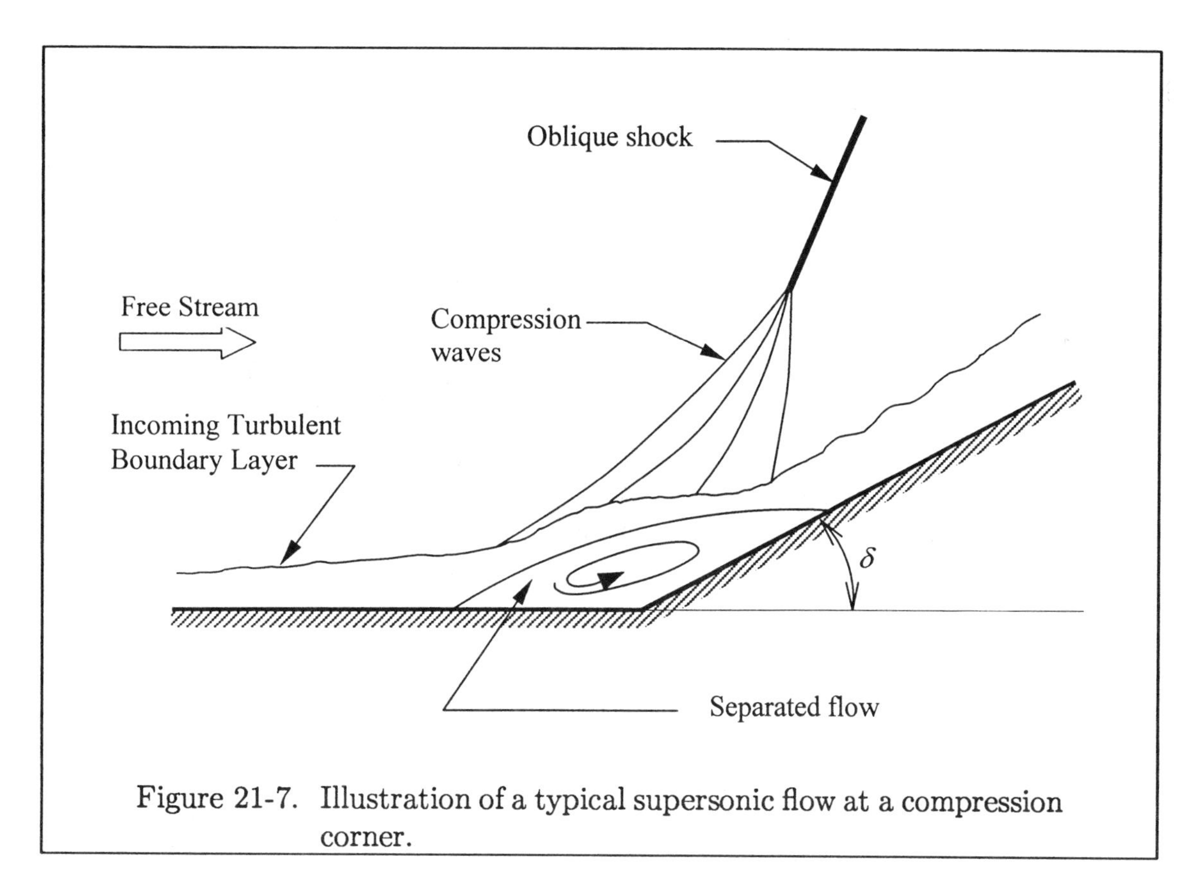

Figure 21-7. Illustration of a typical supersonic flow at a compression corner.

Note that at this stagnation point higher values of pressure and heat transfer are produced. (3) The shock/turbulent boundary layer interaction is highly unsteady. The shock typically oscillates back and forth, creating large pressure fluctuations at the surface. This physical phenomenon has been shown by several researchers, for example, Dolling and his co-workers [21-19 – 21-21].

A detailed flowfield survey for several compression corners have been conducted by Settles [21-23]. The flowfield corresponding to a compression corner of 24° is used in the application presented in this section. The flow conditions are specified as follow, $M_\infty = 2.85(u_\infty = 1875$ ft/sec), $T_\infty = 180$ °R, and a constant wall temperature of $T_w = 498$ °R. The Reynolds number base on δ_0, where $\delta_0 = 0.83$ inches, was 1.33×10^6.

The grid system is composed of 101×81 grid points as shown in Figure 21-8. The grid points clustering near the surface and at the compression corner is enforced. The y^+ for the first grid point off the wall, $(i = 2)$, varied from about 1 at the inlet to about 4 at the downstream of reattachment point. It is important to reiterate that for turbulent flow computations, a y^+ of less than 5, preferably a y^+ of $1 \sim 2$ are necessary. Satisfying this requirement will ensure that at least one grid point is located within the viscous sublayer.

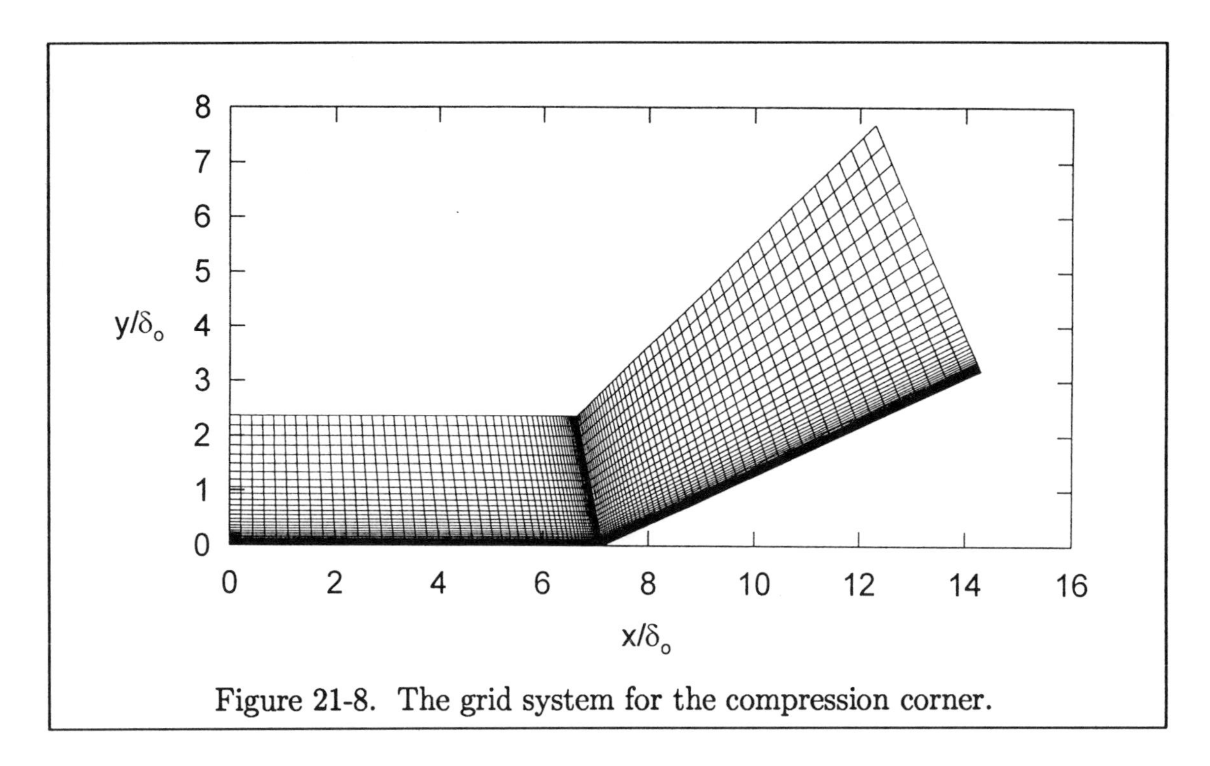

Figure 21-8. The grid system for the compression corner.

The initial conditions at the inlet can be obtained by several methods. For example, a computation can be conducted over a flat plate and at a location where the computed boundary layer thickness matches the experimental value, the data, including the turbulent viscosity distribution, is stored to be used as the inlet conditions.

The no slip boundary condition and the constant wall temperature are imposed at the surface, that is, the lower surface. At the upper boundary, the free stream conditions are specified. Finally, at the downstream boundary, a first-order extrapolation is used. The solution is initiated by imposing the specified flow condition at the inlet over the entire domain.

Comparisons of the computed surface pressure distributions with the experimental data are shown in Figures (21-9) and (21-10). The solutions shown in these figures are obtained using the Spalart-Allmaras one-equation model and the Jones-Launder two-equation k-ϵ model, respectively.

Recall that one of the physical phenomena of the shock/turbulent boundary layer interaction identified previously is the unsteadiness of this interaction. However, the solutions obtained in this example are developed for steady state. There are a couple of reasons for doing so. First, the experimental data used in comparisons are provided as their mean values. Second, and perhaps more important, is the formulations used in the computations. Since an averaging process is used in the development of the Reynolds Average Navier-Stokes equation, the occurrence of any high frequency processes within the domain cannot be predicted. In order to

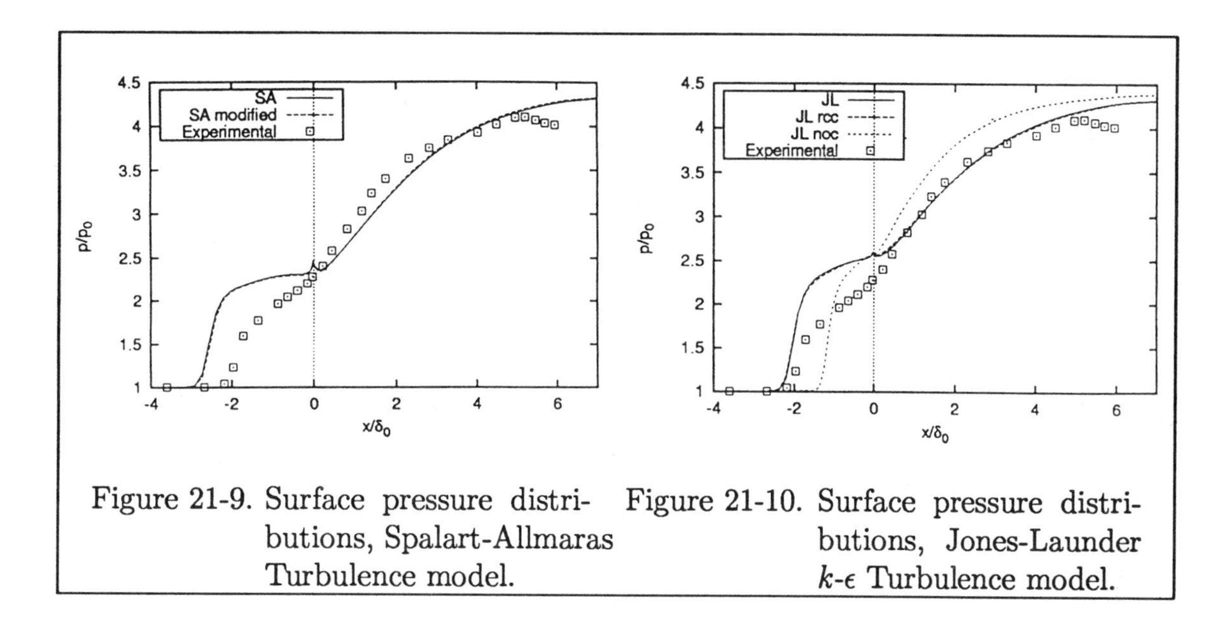

Figure 21-9. Surface pressure distributions, Spalart-Allmaras Turbulence model.

Figure 21-10. Surface pressure distributions, Jones-Launder k-ϵ Turbulence model.

accurately predict the unsteady nature of this problem, large eddy simulation or direct numerical simulation should be used. Nonetheless, the RANS equations complemented with a turbulence model can predict the overall behavior of the flowfield. Comparisons of the velocity profiles at the compression corner and downstream of the corner are shown in Figures (21-11) and (21-12). The computed values compare reasonably well with the experimental data in light of the complexity of the flowfield. Observe that the reverse flow at the corner is predicted fairly well, as shown in Figure (21-11).

It should be noted that, due to the uncertainties in experimentally measured turbulence quantities, the approximations used in the RANS equations, and the uncertainties in turbulence models, a prediction with a range of ± 20 percent can be considered reasonable for complex flows involving turbulence.

21.7.2 Two-Dimensional Base Flow

Near-wake flowfield behind a finite thickness base generated when two streams of turbulent flows merge occurs in several applications. The complexity of flowfield is manifested if the flow is supersonic. In this type of flowfields, the upper and lower streams separate at the corners of the base forming turbulent shear layers. These shear layers interact with the expansion waves near the corners and undergo recompression and reattachment downstream of the base. Near the base, there is a low speed-recirculating region approximately at a constant pressure bounded by the shear layers. The base flowfield is illustrated in Figure 21-13.

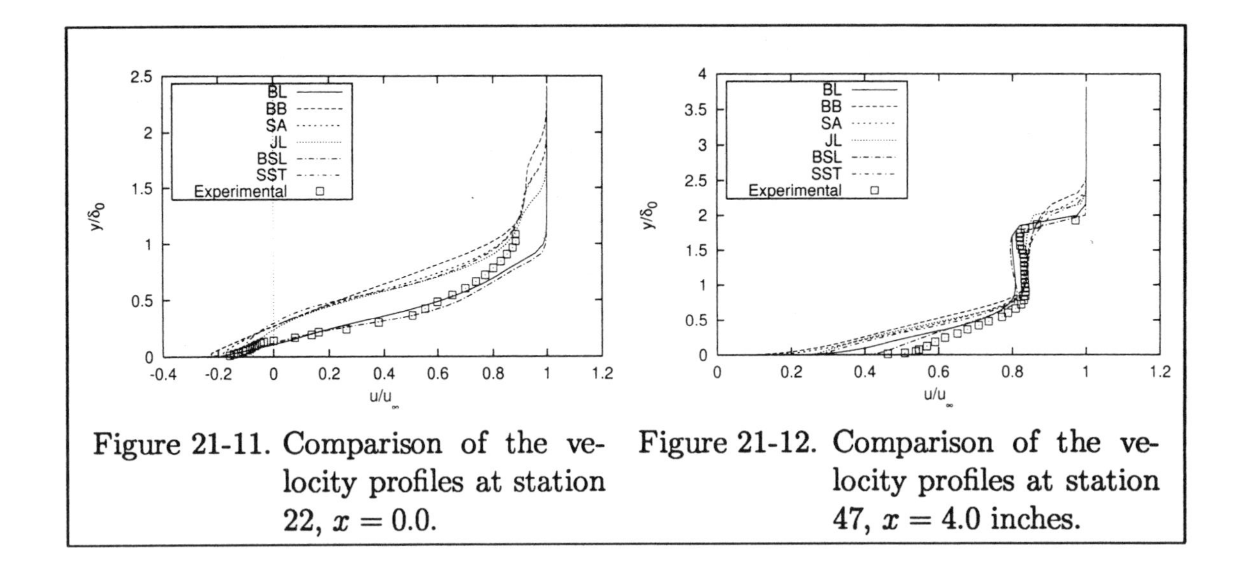

Figure 21-11. Comparison of the velocity profiles at station 22, $x = 0.0$.

Figure 21-12. Comparison of the velocity profiles at station 47, $x = 4.0$ inches.

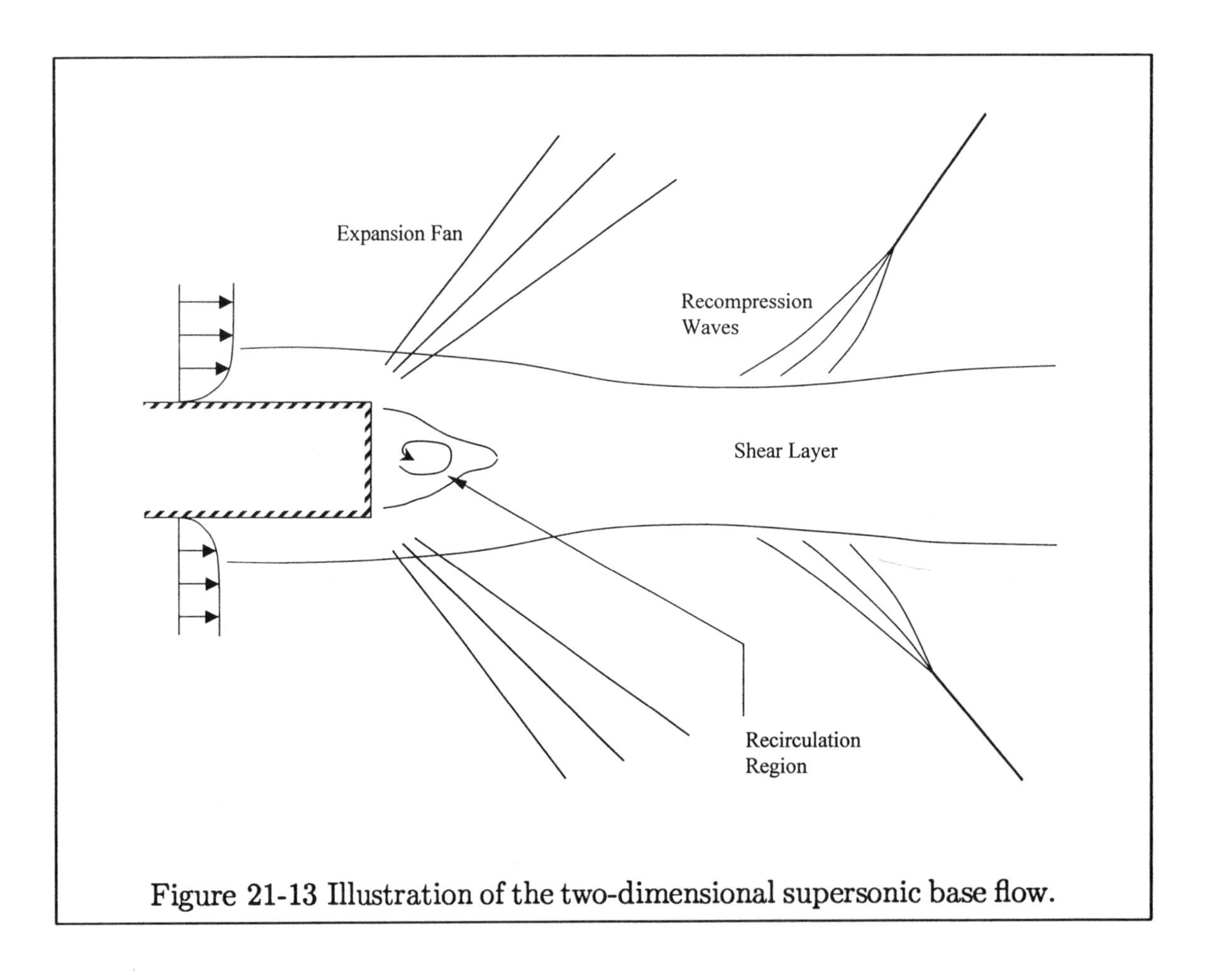

Figure 21-13 Illustration of the two-dimensional supersonic base flow.

The objectives of this exercise are to numerically simulate this complex flowfield and, in the process, to evaluate the performance of several turbulence models discussed previously.

The flow properties at the upper and lower streams are designated by subscripts 1 and 2, respectively, and they are specified as

$$M_1 = 2.56, \quad u_1 = 584.13 \text{ m/sec}, \quad p_1 = 27,566 \text{ N/m}^2$$

and

$$M_2 = 2.05, \quad u_2 = 523.83 \text{ m/sec}, \quad p_2 = 61,110 \text{ N/m2}$$

The corresponding total pressure and total temperature are 517KP and 299 K, respectively. The properties of stream 1 are used to nondimensionalize all the flow properties.

The domain of solution is shown in Figure 21-14, where the origin of the coordinate system is located at the upper corner, as illustrated.

Computations are performed on a multi-block grid system where the domain is divided into three blocks, as shown in Figure 21-15. The governing equations are solved on each block by communicating the solutions at the connection boundaries of blocks. Since this approach has not been previously introduced, some deliberation is warranted.

Consider a domain shown in Figure 21-16. For simplicity assume that it is required to decompose the domain into two blocks, as shown in Figure 21-17, where transfer of data between the two blocks is accomplished through the connection boundary. The overlapping lines CD and EF are used as the connection boundary between blocks 1 and 2. Note that line EF is identified by $i = IM1$ for block 1, and $i = 2$ for block 2. Similarly, line CD is identified as $i = IM1 - 1$ for block 1 and line $i = 1$ for block 2.

The solution procedure consists of two steps at each time level. Solution begins in block 1 where, for the first time, the specified initial condition is used to initiate the solution. Subsequently, solution proceeds to block 2 where the data along line CD $(i = IM1 - 1)$ of the block 1 which has just been completed are used as the boundary data along $i = 1$ of block 2. That is, the boundary values are set according to

$$(\rho_{1,j})\text{Block 2} = (\rho_{IM1-1,j})\text{Block 1}$$

$$(u_{1,j})\text{Block 2} = (u_{IM1-1,j})\text{Block 1}$$

$$(v_{1,j})\text{Block 2} = (v_{IM1-1,j})\text{Block 1}$$

$$(e_{t_{1,j}})\text{Block 2} = (e_{t_{IM1-1,j}})\text{Block 1}$$

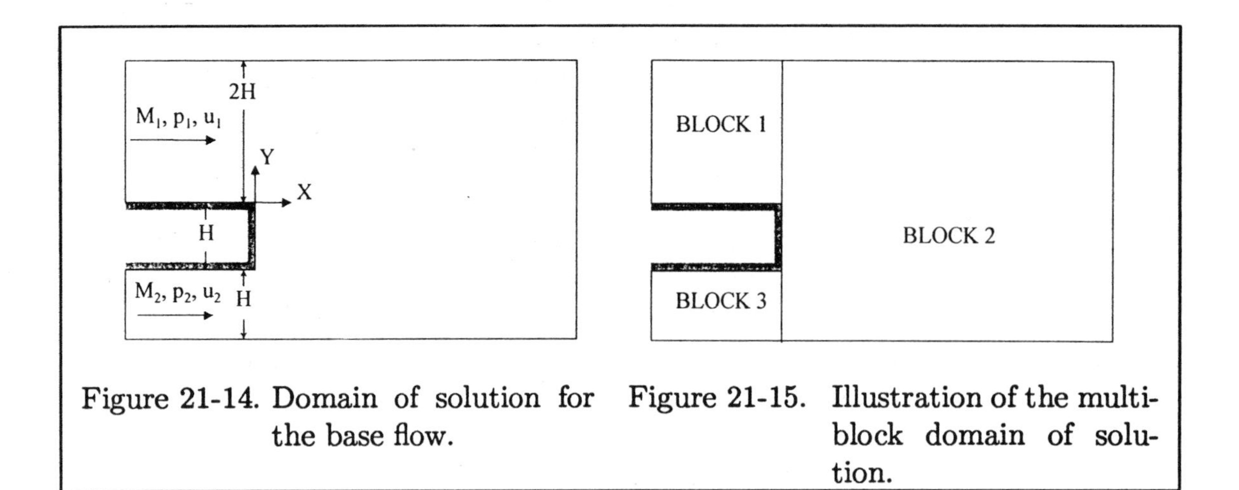

Figure 21-14. Domain of solution for the base flow.

Figure 21-15. Illustration of the multi-block domain of solution.

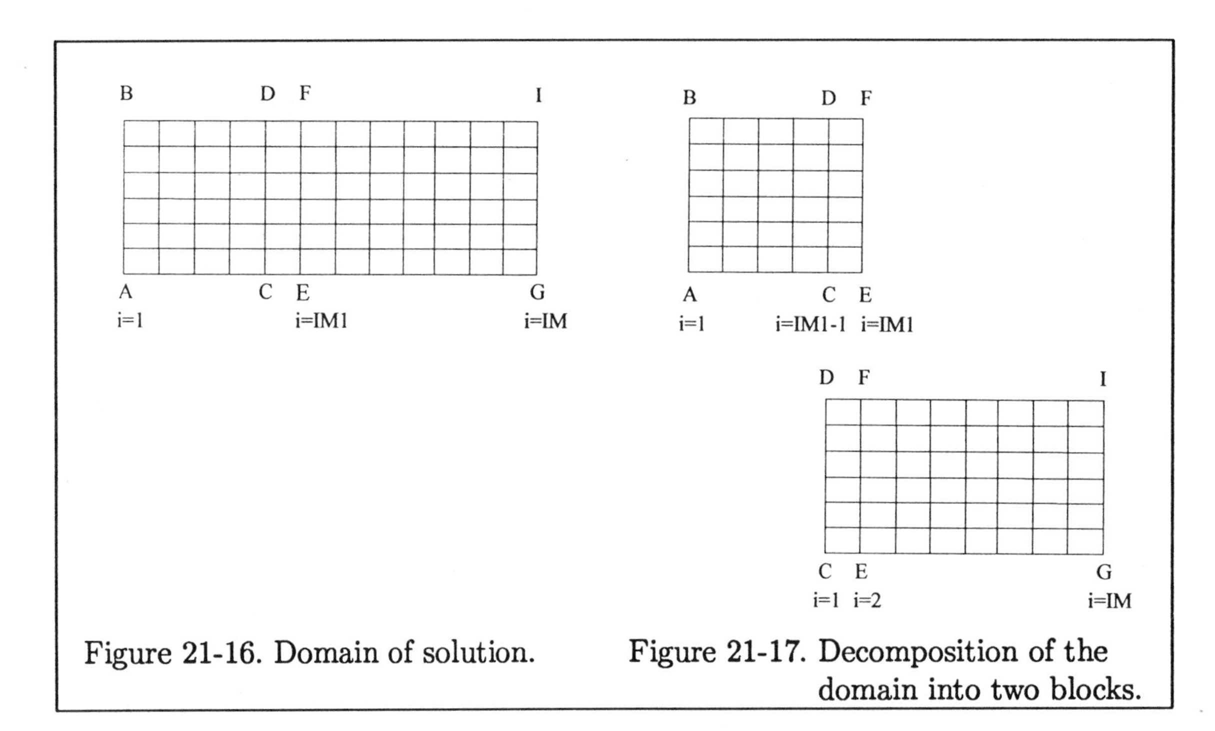

Figure 21-16. Domain of solution.

Figure 21-17. Decomposition of the domain into two blocks.

Once the solution for block 2 is completed, the computed values of the variables along the line EF of block 2 are used as the boundary values along line $EF(i = IM1)$ of block 1, and solution of block 1 is initiated for the next time level. The specification of boundary condition from block 2 to block 1 is set according to

$$(\rho_{IM1,j})\text{Block 1} = (\rho_{2,j})\text{Block 2}$$

$$(u_{IM1,j})\text{Block 1} = (u_{2,j})\text{Block 2}$$

$$(v_{IM1,j})\text{Block 1} = (v_{2,j})\text{Block 2}$$

$$(e_{t_{IM1,j}})\text{Block 1} = (e_{t_{2,j}})\text{Block 2}$$

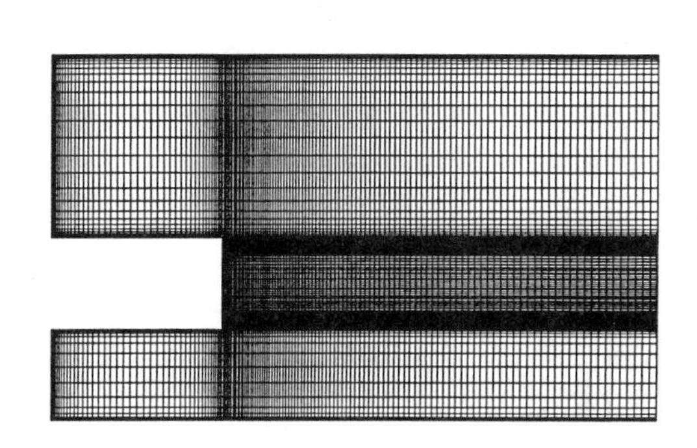

Figure 21-18. Initial rectangular grid block 1: 44×32, block 2: 100×131, block 3: 44×18.

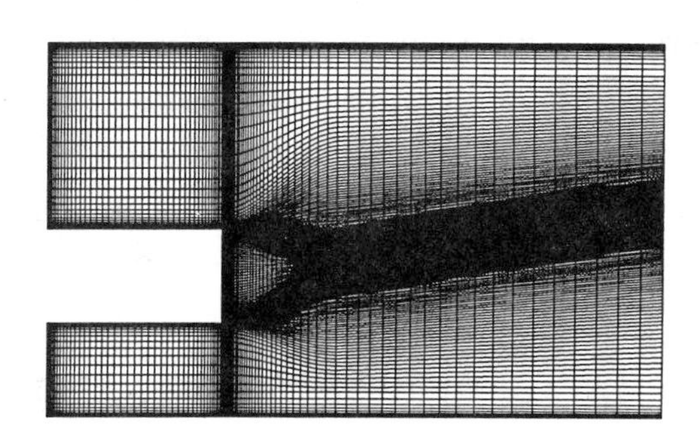

Figure 21-19. Adapted grid system, block 1: 44×70, block 2: 100×245, block 3: 44×50.

Computations for the base flow are performed on the rectangular multi-block grid system shown in Figure 21-18. This initial computation is performed to obtain an estimate of the shear layer in order to generate a more appropriate grid where there is a clustering of grid in the shear layer region, and, furthermore, the grid lines are more or less aligned along the shear layer. Thus, once a solution on the grid system shown in Figure 21-18 is obtained, a second grid based on the flow data is generated which is shown in Figure 21-19. Subsequently, the solution from the initial rectangular grid is mapped onto the new grid system and used as the initial condition. Turbulent flow computations are initiated on this grid. Transition to turbulence is specified on the upper and lower surfaces in blocks 1 and 3 at a distance of half base thickness downstream of the inflow. Observe that there is clustering of grid points along the upper and lower walls and at the inlet location in addition to grid clustering along the shear layer.

As the solution develops, the grid system may be updated in order to better align the grid line along the shear layer and to impose more efficiently the grid clustering. The inlet flow conditions are generated in a similar approach, as described in the previous example.

Computed streamwise mean velocity profiles obtained by the Baldwin-Barth one-equation model, the Spalart-Allmaras one-equation model, and the Baseline k-ϵ/k-ω two-equation model with compressibility correction are shown in Figures 21-20 through 21-22. The axial locations in Figures 21-20 through 21-22 are at x = 5mm, 25mm, and 55 mm, respectively. Observation of the experimental data indicates that the axial locations of x = 5mm and 25mm approximately correspond to the separation/recirculation region and the shear layer mixing region, where an axial location of x = 55mm corresponds to the recompression/reattachment region.

The streamwise velocity profiles obtained by the baseline two-equation turbulence model with compressibility correction provide the best prediction, as compared to the experimental data. Comparison of the baseline pressure is shown in Figure 21-23. The pressure values obtained from the baseline two-equation model with compressibility correction and the Baldwin-Barth model compare well with the experimental data, whereas the pressure values predicted by the baseline and Spalart-Allmaras models are much lower.

21.7.3 Axisymmetric, Supersonic, Turbulent Exhaust Flow

A typical supersonic exhaust flow includes viscous/inviscid interaction, turbulence, and possibly it may include thermochemistry. The boundary layers developed in the interior and/or exterior walls of the nozzle produce shear layers at the exit plane. A complex series of expansion waves and compression waves are generated which interact with the shear layer. As a consequence, the so-called barrel shocks are formed. Shock reflection from the axis of symmetry may occur in the form of either a Mach Reflection or a regular reflection. A typical flowfield is illustrated in Figure (21-24). The plume structure is further influenced by the external flow conditions, which may be subsonic or supersonic. Other significant physical factors which influence the plume flowfield are turbulence and thermochemistry. Typically the shear layer is turbulent, and, therefore, appropriate turbulence models must be considered in flow simulations. Furthermore, due to large variations in the temperature field within the plume, a proper chemistry model should be utilized to capture the physics of the problem. In fact, the temperature behind the Mach disc could reach as high as the stagnation temperature. Since the primary objective in this section is to investigate solutions of several turbulence models without any added complexity, the effect of chemistry is not included and the perfect gas model is used. Indeed, for the example problem considered, temperatures do not reach values for which chemistry consideration would be warranted.

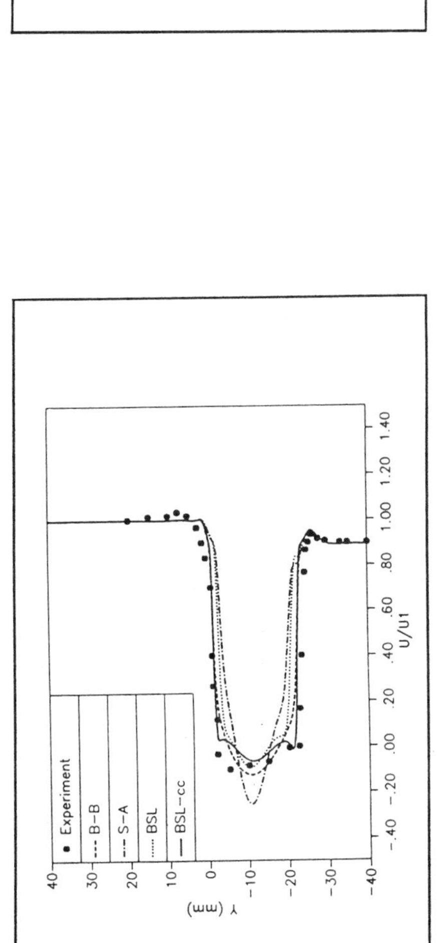

Figure 21-20. Comparison of the streamwise mean velocity profiles at $x = 5$mm.

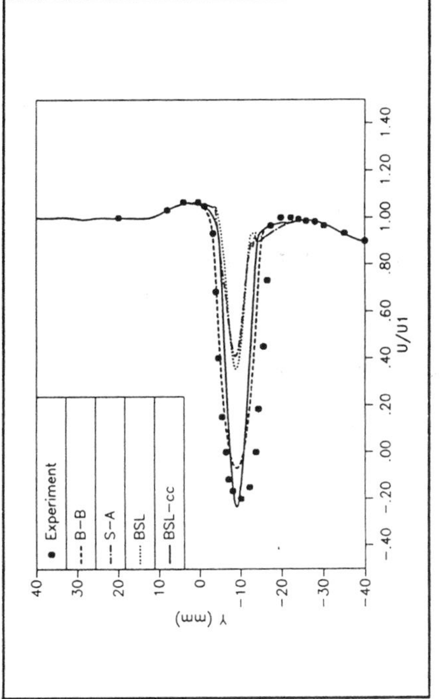

Figure 21-21. Comparison of the streamwise mean velocity profiles at $x = 25$mm.

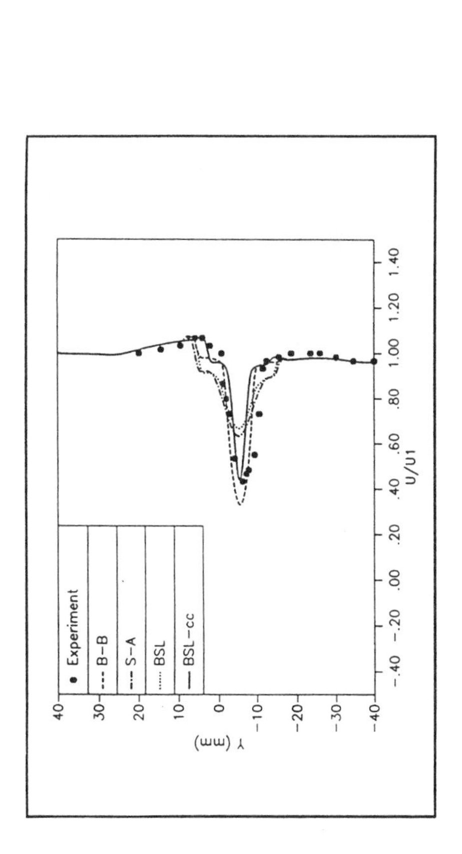

Figure 21-22. Comparison of the streamwise mean velocity profiles at $x = 55$mm.

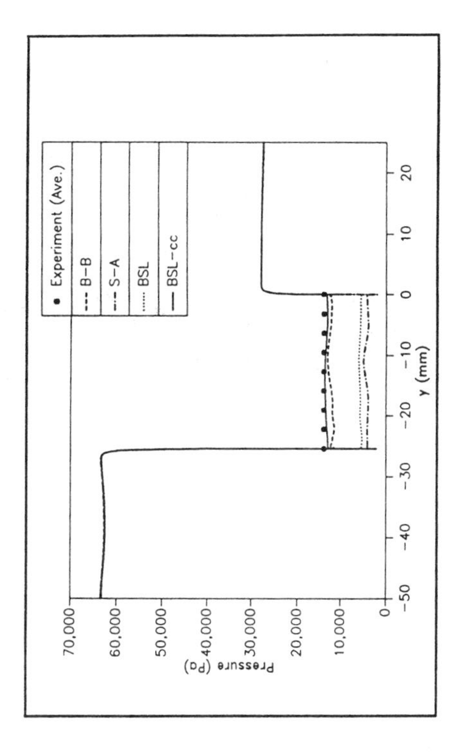

Figure21-23. Comparison of the pressure distribution at the base.

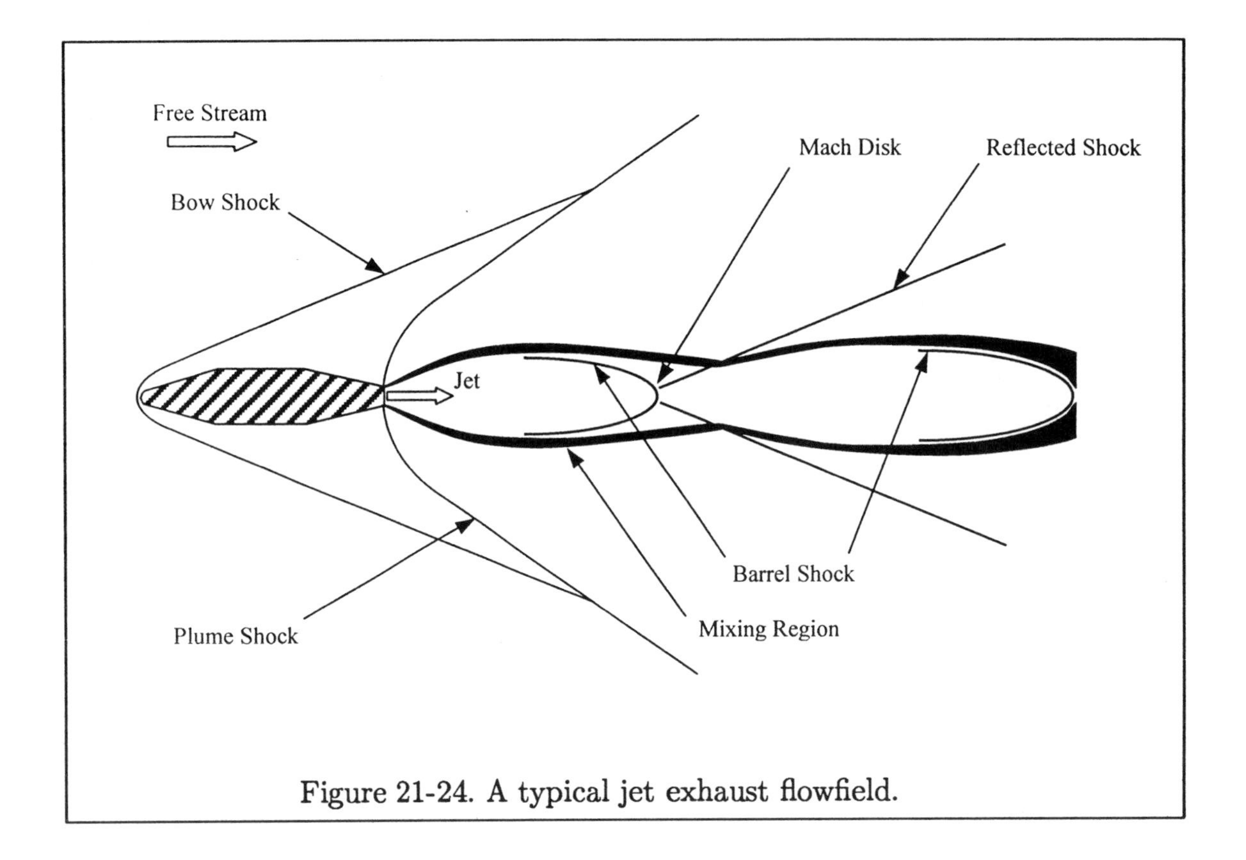

Figure 21-24. A typical jet exhaust flowfield.

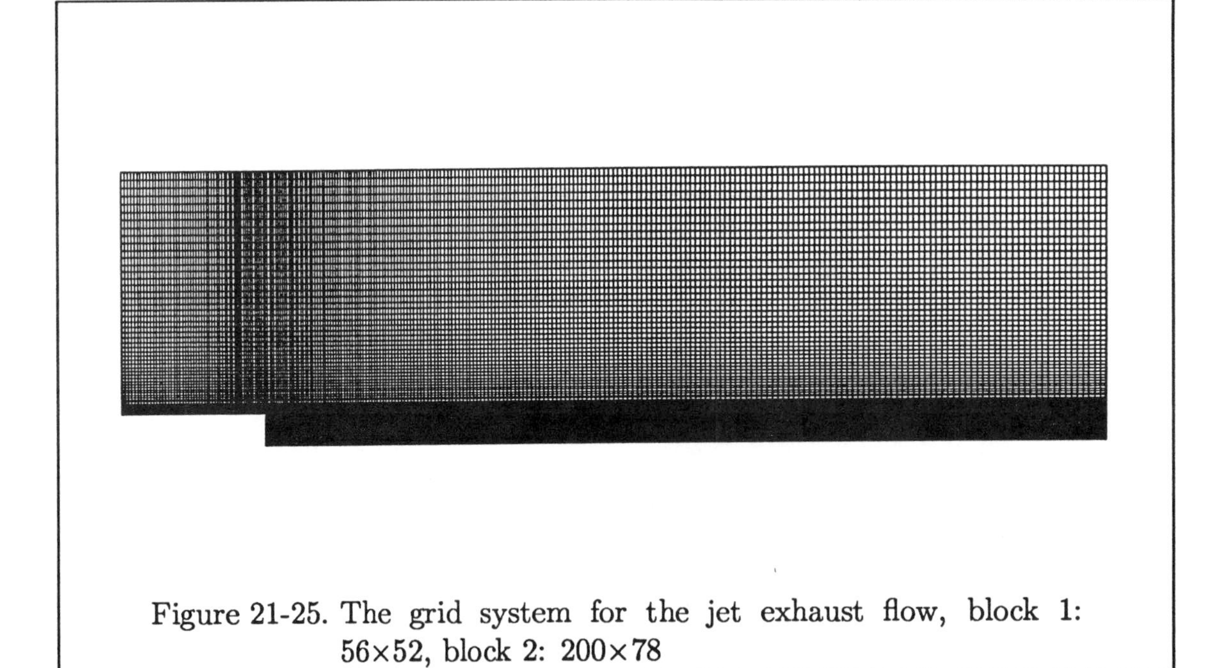

Figure 21-25. The grid system for the jet exhaust flow, block 1: 56×52, block 2: 200×78

Consider now the exhaust flow of an underexpanded nozzle at Mach 2 into a quiescent ambient with a pressure ratio of $P_j/P_b = 1.45$, where P_j is the jet exit pressure and P_b is the back pressure. Furthermore, the temperature ratio is given as $T_j/T_b = 1.0$. The freestream pressure and temperature are 101,325 N/m^2 and 293K, respectively. The jet exit radius R_j is 2.54cm.

A grid system composed of two blocks, as shown in Figure 21-25, is used for computations. The grid system includes 56×52 and 200×78 grid points for blocks 1 and 2, respectively. Due to symmetry, only the upper half of the domain is considered in computation. Block 1 extends upstream from the jet exit plane $2.5R_j$, whereas block 2 extends $15R_j$ downstream from the jet exit plane. The outer boundary is set at $9R_j$ from the axis of symmetry. The origin of the coordinate system is set at the jet exit plane on the axis of symmetry.

The boundary condition at the jet exit plane is specified, based on the experimentally measured velocity profile. In the course of investigation, it was concluded that the solution is sensitive to specification of boundary conditions set at the freestream. In fact, depending on how the boundary conditions were imposed, it was necessary in some cases to specify a small external coflow in the order of $M_\infty = 0.01$ or larger in order to obtain a stable solution. A detailed investigation of boundary conditions for the jet exhaust flow is provided by Hoffmann, et al., [21-23].

Turbulent flow computations are initiated by setting onset of transition at three grid points downstream of the jet exit plane. The transition is simulated by phasing in the production term in the turbulence model over several (e.g., three) grid points.

The center line pressure distribution obtained with the Baldwin-Barth model and a modified Baldwin-Barth model are shown in Figure 21-26, where the numerical solutions are compared to the experimental data of Seiner and Norum [21-24]. The Baldwin-Barth model used in this example corresponds to Equation (21-77), with $c_{\epsilon 1} = 1.2$. The modified version, which includes axisymmetric correction, is again Equation (21-77) with simply increasing the value of $c_{\epsilon 1}$ to 1.47.

Solutions by Spalart-Allmaras model are shown in Figure 21-27. The Spalart-Allmaras one-equation model is given by Equation (21-124) where c_{b1} is specified as 0.1355. Observe that this model produces an excessive amount of turbulent dissipation, and, as a result, the shock cells are dissipated at around x/d of about 10. The modified Spalart-Allmaras model is the same as given by Equation (21-124), except a value of 0.06 is used for c_{b1}. Finally, the solutions by the baseline k-ϵ/k-ω two-equation model are shown in Figure 21-28. Again, as in the case of the Spalart-Allmaras model, the original baseline model produces excessive turbulent dissipation. Terms are included for compressibility correction (BSL-cc) with $\alpha_1 =$ 1.0, $\alpha_2 = 0.4$, and $\alpha_3 = 0.2$, and compressibility and axisymmetric corrections (BSLcca) with $\alpha_1 = 1.0$, $\alpha_2 = 0.4$, $\alpha_3 = 0.2$, and $\alpha_2 = 0.6$. The best solution is obtained when both compressibility and axisymmetric correction terms are included.

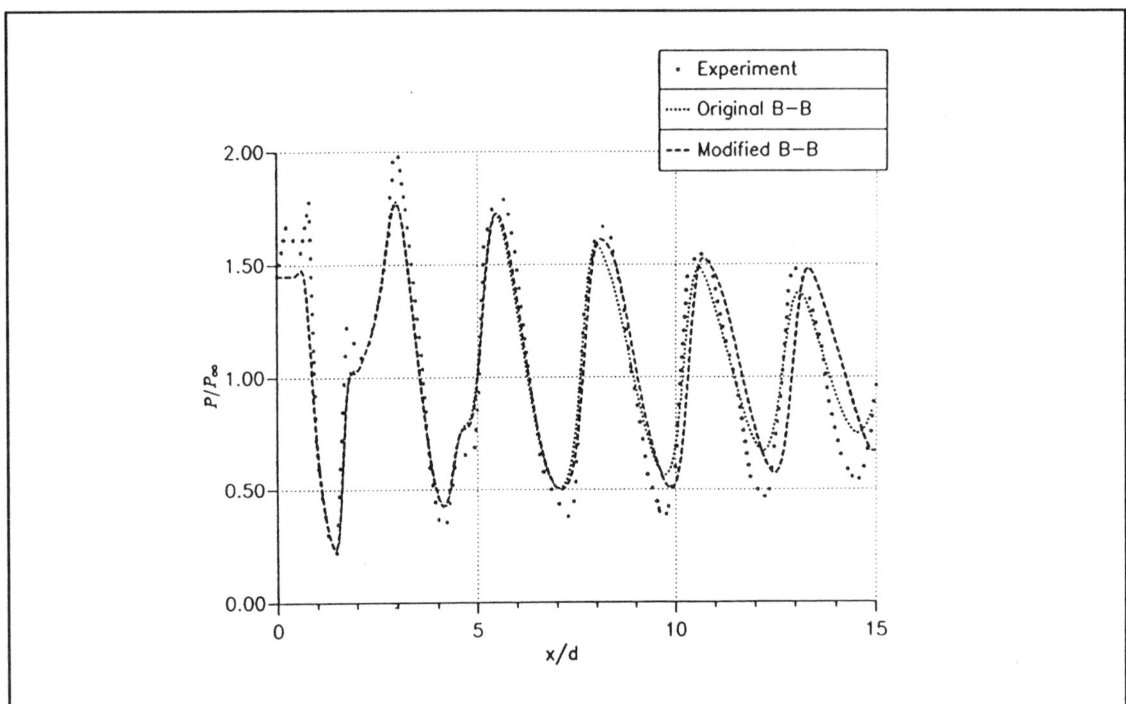

Figure 21-26. Comparison of the pressure distributions along the jet centerline, Baldwin-Barth model.

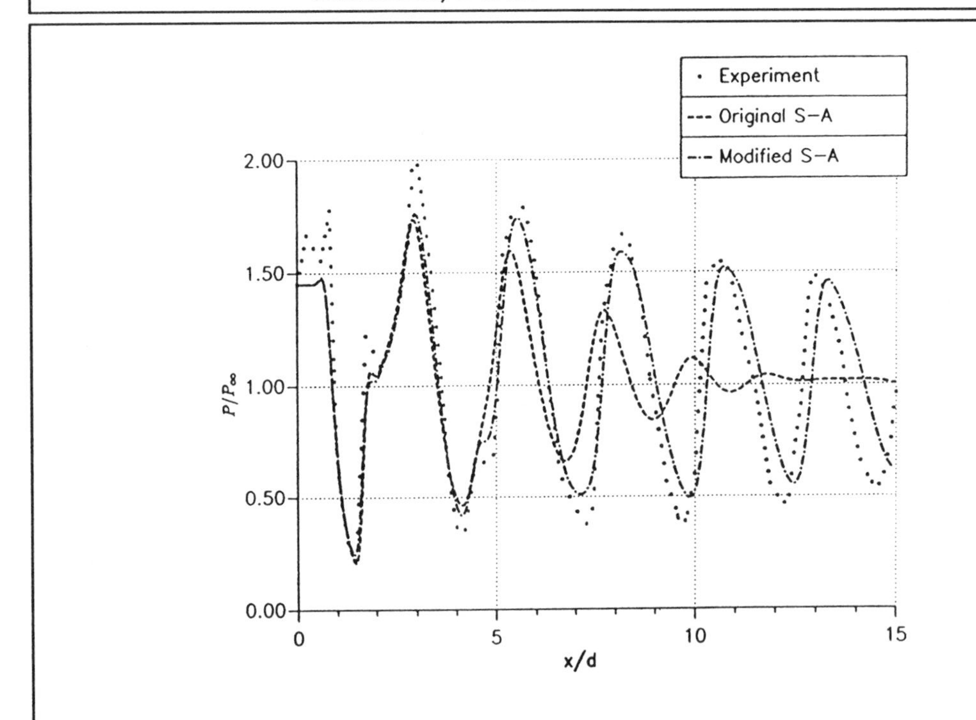

Figure 21-27. Comparison of the pressure distributions along the jet centerline, Spalart-Allmaras model.

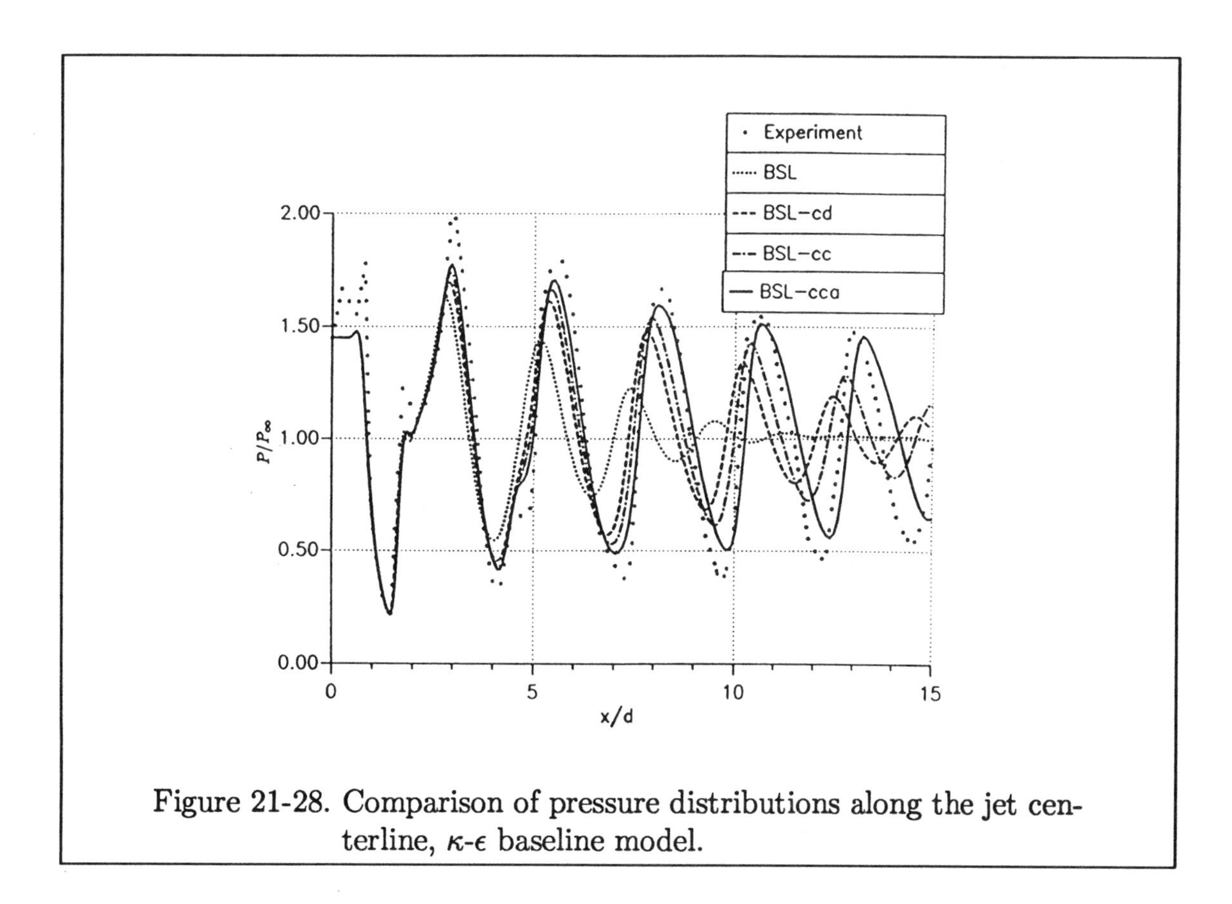

Figure 21-28. Comparison of pressure distributions along the jet centerline, κ-ϵ baseline model.

21.8 Favre-Averaged Navier-Stokes Equations

Previously in Section 21.3.1, the Reynolds-averaged Navier-Stokes equations, based on time averaging, were developed. One of the main assumptions implemented in the development of RANS is the Morkovin's hypothesis which allows the elimination of terms involving density fluctuations. Recall that Morkovin's hypothesis states that the structure of turbulence is negligibly affected by compressibility up to Mach number of about 5 and remains similar to that of incompressible flow. To remove this assumption and to increase the range of applications to high speed compressible flows, the density fluctuating correlations such as $\overline{\rho' u'}$, $\overline{\rho' v'}$ are retained. The resulting equations are known as the *Favre-averaged Navier-Stokes (FANS) equations* which are also referred to as the *mass-averaged Navier-Stokes equations.* Thus, the terms *Favre-averaged* and *mass-averaged* are used interchangeably. The objective of this section is to review the development of FANS equations.

A mass-(or Favre-)averaged quantity denoted by a tilde over the variable is define as

$$\tilde{f} = \frac{1}{\bar{\rho}} \int_{t_o}^{t_o+\Delta t} \rho f \, dt = \frac{\overline{\rho f}}{\bar{\rho}} \tag{21-302}$$

where, as before, a bar over a variable indicates a Reynolds-averaged quantity. It is important to note that Favre-averaging is a mathematical concept introduced to simplify the equations, and it is not a physical phenomenon.

Relation (21-302) can be written as

$$\bar{\rho}\,\tilde{f} = \overline{\rho f} = \overline{(\bar{\rho} + \rho')\,(\bar{f} + f')} = \bar{\rho}\,\bar{f} + \overline{\rho' f'} \tag{21-303}$$

Any quantity f can be decomposed as

$$f = \tilde{f} + f'' \tag{21-304}$$

where $\tilde{f}$ and f'' are the Favre mean and Favre fluctuating part of the turbulent motion. Relation (21-304) can be multiplied by ρ and time-averaged to yield

$$\overline{\rho f} = \bar{\rho}\,\tilde{f} + \overline{\rho f''} \tag{21-305}$$

Recall that from relation (21-303)

$$\overline{\rho f} = \bar{\rho}\,\tilde{f} \tag{21-306}$$

and therefore it follows that

$$\overline{\rho f''} = 0 \tag{21-307}$$

Now, from definition (21-304), one has

$$f'' = f - \tilde{f} \tag{21-308}$$

and from (21-303)

$$\tilde{f} = \bar{f} + \frac{\overline{\rho' f'}}{\bar{\rho}} \tag{21-309}$$

The combination of relations (21-308) and (21-309) yields

$$f'' = f - \tilde{f} = f - \bar{f} - \frac{\overline{\rho' f'}}{\bar{\rho}} = f' - \frac{\overline{\rho' f'}}{\bar{\rho}} \tag{21-310}$$

where relation (21-19) is also used. Subsequently, the time-averaging of (21-310) provides

$$\overline{f''} = \overline{f'} - \frac{\overline{\rho' f'}}{\bar{\rho}}$$

But since, according to (21-21), $\overline{f'} = 0$, then

$$\overline{f''} = -\frac{\overline{\rho' f'}}{\bar{\rho}} \neq 0 \tag{21-311}$$

Now, the rules of Favre-averaging are summarized as follows

$$\bar{\tilde{f}} = \tilde{f} \qquad (21\text{-}312)$$

$$\overline{\rho f''} = 0 \qquad (21\text{-}313)$$

$$\overline{f''} = -\frac{\overline{\rho' f'}}{\bar{\rho}} \qquad (21\text{-}314)$$

$$\overline{\rho f g} = \bar{\rho}\tilde{f}\tilde{g} + \overline{\rho f'' g''} \qquad (21\text{-}315)$$

With the definitions and rules of Favre-averaging established, consider the time- or Reynolds-averaged continuity equation (for a two-dimensional flow) as given by Equation (21-23), that is,

$$\frac{\partial \bar{\rho}}{\partial t} + \frac{\partial}{\partial x}\left(\bar{\rho}\,\bar{u} + \overline{\rho' u'}\right) + \frac{\partial}{\partial y}\left(\bar{\rho}\,\bar{v} + \overline{\rho' v'}\right) = 0 \qquad (21\text{-}316)$$

Using relation (21-303), the Favre-averaged continuity equation now can be written as

$$\frac{\partial \bar{\rho}}{\partial t} + \frac{\partial}{\partial x}(\bar{\rho}\,\tilde{u}) + \frac{\partial}{\partial y}(\bar{\rho}\,\tilde{v}) = 0 \qquad (21\text{-}317)$$

Observe that the Favre-averaged continuity equation retains the same form as the Reynolds-averaged continuity equation, and, in fact, the same form as the original continuity equation. This is precisely the reason for the definition of mass-averaged quantities, as given by (21-303). That is, the mass-averaged quantities are defined in such a way as to write the resulting equation in a similar form as the original equations, while retaining the effect of density fluctuations. Furthermore, this is the reason why it was stated previously that the concept of Favre-averaging is mathematical.

Now consider the x-component of the momentum equation for a two-dimensional flow given by

$$\frac{\partial}{\partial t}(\rho u) + \frac{\partial}{\partial x}(\rho u^2) + \frac{\partial}{\partial y}(\rho u v) = -\frac{\partial p}{\partial x} + \frac{\partial \tau_{xx}}{\partial x} + \frac{\partial \tau_{xy}}{\partial y} \qquad (21\text{-}318)$$

For simplicity, each term will be considered separately. The procedure, as illustrated previously, begins with time-averaging, and subsequently the Favre-averaging concept is applied. Now the first term is written as

$$\frac{\partial}{\partial t}(\overline{\rho u}) = \frac{\partial}{\partial t}\left[\overline{(\bar{\rho} + \rho')(\bar{u} + u')}\right] = \frac{\partial}{\partial t}(\bar{\rho}\,\bar{u} + \overline{\rho' u'}) = \frac{\partial}{\partial t}(\bar{\rho}\,\tilde{u})$$

Similarly, the remaining terms are modified and given by the following

$$\frac{\partial}{\partial x}(\overline{\rho u^2}) = \frac{\partial}{\partial x}(\bar{\rho}\,\tilde{u}^2 + \overline{\rho u''^2})$$

$$\frac{\partial}{\partial y}(\overline{\rho u v}) = \frac{\partial}{\partial y}\left(\bar{\rho}\,\tilde{u}\,\tilde{v} + \overline{\rho\, u''\, v''}\right)$$

$$\bar{\tau}_{xx} = \tilde{\mu}\left(\frac{4}{3}\frac{\partial \tilde{u}}{\partial x} - \frac{2}{3}\frac{\partial \tilde{v}}{\partial y}\right)$$

and

$$\bar{\tau}_{xy} = \tilde{\mu}\left(\frac{\partial \tilde{u}}{\partial y} + \frac{\partial \tilde{v}}{\partial x}\right) \tag{21-319}$$

Therefore, the x-component of the momentum equation, given by (21-318), is written in a mass-averaged form as

$$\frac{\partial}{\partial t}(\bar{\rho}\,\tilde{u}) + \frac{\partial}{\partial x}(\bar{\rho}\,\tilde{u}^2) + \frac{\partial}{\partial y}(\bar{\rho}\,\tilde{u}\,\tilde{v}) = -\frac{\partial \bar{p}}{\partial x} + \frac{\partial}{\partial x}(\bar{\tau}_{xx} - \overline{\rho\, u''^2}) + \frac{\partial}{\partial y}(\bar{\tau}_{xy} - \overline{\rho\, u''\, v''}) \tag{21-320}$$

Note that the density in $\overline{\rho\, u''^2}$ and $\overline{\rho\, u''\, v''}$ is the instantaneous density, and that these terms represent Reynolds stresses.

Similarly, the y-component of the momentum equation is written as

$$\frac{\partial}{\partial t}(\bar{\rho}\,\tilde{v}) + \frac{\partial}{\partial x}(\bar{\rho}\,\tilde{u}\,\tilde{v}) + \frac{\partial}{\partial y}(\bar{\rho}\,\tilde{v}^2) = -\frac{\partial \bar{p}}{\partial y} + \frac{\partial}{\partial x}(\bar{\tau}_{xy} - \overline{\rho\, u''\, v''}) + \frac{\partial}{\partial y}(\bar{\tau}_{yy} - \overline{\rho\, v''^2}) \tag{21-321}$$

The energy equation in terms of the total energy per unit mass expressed in Cartesian coordinates for two-dimensional problems is given by

$$\frac{\partial}{\partial t}(\rho\, e_t) + \frac{\partial}{\partial x}(\rho\, u\, e_t + p u) + \frac{\partial}{\partial y}(\rho\, v\, e_t + p v) =$$

$$\frac{\partial}{\partial x}(u\,\tau_{xx} + v\,\tau_{xy} - q_x) + \frac{\partial}{\partial y}(u\,\tau_{xy} + v\,\tau_{yy} - q_y) \tag{21-322}$$

Again, considering the modification of each term separately, the first term is

$$\frac{\partial}{\partial t}(\rho\, e_t) = \frac{\partial}{\partial t}\left[\overline{(\bar{\rho} + \rho')\,(\bar{e}_t + e_t')}\right] = \frac{\partial}{\partial t}(\bar{\rho}\,\bar{e}_t + \overline{\rho'\, e_t'}) = \frac{\partial}{\partial t}(\bar{\rho}\,\tilde{e}_t)$$

The term $\bar{\rho}\,\tilde{e}_t$ can be rewritten as

$$\bar{\rho}\,\tilde{e}_t = \overline{\rho\, e_t} = \overline{\rho\, c_v T} + \frac{1}{2}\overline{\rho (u^2 + v^2)} \tag{21-323}$$

where the following relation is used

$$e_t = e + \frac{1}{2}V^2 = c_v T + \frac{1}{2}(u^2 + v^2)$$

Observe that, in the equation above, the potential energy has been dropped.

Now the instantaneous terms in Equation (21-323) are replaced by their mean and fluctuating values, thus,

$$\overline{\rho c_v T} + \frac{1}{2}\overline{\rho(u^2+v^2)} = c_v\,\overline{(\bar{\rho}+e')\,(\bar{T}+T')} + \frac{1}{2}\overline{\rho u^2} + \frac{1}{2}\overline{\rho v^2}$$

$$= c_v(\bar{\rho}\bar{T} + \overline{e'T'}) + \frac{1}{2}\overline{\rho u^2} + \frac{1}{2}\overline{\rho v^2}$$

Using relations (21-309) and (21-315), one can now write

$$\bar{\rho}\,\tilde{e}_t = c_v(\bar{\rho}\tilde{T}) + \frac{1}{2}\bar{\rho}\,\tilde{u}^2 + \frac{1}{2}\overline{\rho\,u''^2} + \frac{1}{2}\bar{\rho}\,\tilde{v}^2 + \frac{1}{2}\overline{\rho\,v''^2}$$

or

$$\bar{\rho}\,\tilde{e}_t = c_v\,\bar{\rho}\,\tilde{T} + \frac{1}{2}\bar{\rho}\,(\tilde{u}^2+\tilde{v}^2) + \frac{1}{2}\,(\overline{\rho u''^2} + \overline{\rho\,v''^2}) \tag{21-324}$$

The last term in relation (21-324) is defined as the *turbulence kinetic energy* (per unit mass) as follows

$$\bar{\rho}k = \frac{1}{2}\overline{\rho\,(u''^2+v''^2)}$$

or

$$k = \frac{1}{2}\frac{\overline{\rho\,(u''^2+v''^2)}}{\bar{\rho}} \tag{21-325}$$

It is important to note that the turbulence kinetic energy defined in the Favre-averaged formulation is different from the turbulence kinetic energy defined in the Reynolds-averaged formulation given by (21-75) as

$$k = \frac{1}{2}\,\overline{(u'^2+v'^2)} \tag{21-326}$$

which is essentially based on incompressible flow considerations. Now relation (21-292) can be written as

$$\bar{\rho}\,\tilde{e}_t = c_v\,\bar{\rho}\,\tilde{T} + \frac{1}{2}\bar{\rho}\,(\tilde{u}^2+\tilde{v}^2) + \bar{\rho}\,k \tag{21-327}$$

The second term of the energy equation (21-322) is considered next, which is rewritten as

$$\rho\,u\,e_t + p\,u = (\rho\,e_t + p)\,u = \left[\rho\,c_v\,T + \frac{1}{2}\rho\,(u^2+v^2) + p\right]u \tag{21-328}$$

Using the perfect gas equation of state, one may write

$$\rho\,c_v\,T + p = \rho\,c_v\,T + \rho RT = \rho\,T(c_v+R) = \rho\,c_p\,T$$

Thus, (21-328) is written as

$$\left[\rho\, c_p T + \frac{1}{2}\rho\,(u^2+v^2)\right]\, u = c_p\,\rho\, u\, T + \frac{1}{2}\rho\, u\,(u^2+v^2) \tag{21-329}$$

which can be time-averaged to provide

$$\overline{\rho\, u\, T} = \bar{\rho}\,\tilde{u}\,\tilde{T} + \overline{\rho\, u''\, T''} \tag{21-330}$$

according to (21-315), and

$$\begin{aligned}
\frac{1}{2}\overline{\rho\, u\,(u^2+v^2)} &= \frac{1}{2}\overline{\rho\, u^3} + \frac{1}{2}\overline{\rho\, u\, v^2} \\
&= \frac{1}{2}\bar{\rho}\,\tilde{u}\,(\tilde{u}^2+\tilde{v}^2) + \tilde{u}\left[\frac{1}{2}\overline{\rho\,(u''^2+v''^2)}\right] + \tilde{u}\overline{\rho\, u''^2} \\
&\quad + \tilde{v}\overline{\rho\, u''v''} + \frac{1}{2}\left(\overline{\rho\, u''^3} + \overline{\rho\, u''v''^2}\right) \\
&= \frac{1}{2}\bar{\rho}\,\tilde{u}\,(\tilde{u}^2+\tilde{v}^2) + \bar{\rho}\, k\,\tilde{u} + \tilde{u}\,\overline{\rho\, u''^2} + \tilde{v}\,\overline{\rho\, u''v''} \\
&\quad + \frac{1}{2}\left(\overline{\rho\, u''^3} + \overline{\rho\, u''v''^2}\right)
\end{aligned}$$

Finally, the second term is written as

$$\begin{aligned}
\overline{(\rho\, e_t + p)\, u} &= \bar{\rho}\,\tilde{u}\left[c_p\tilde{T} + \frac{1}{2}(\tilde{u}^2+\tilde{v}^2) + k\right] + c_p\,\overline{\rho\, u''T''} \\
&\quad + \tilde{u}\,\overline{\rho u''^2} + \tilde{v}\,\overline{\rho\, u''v''} + \frac{1}{2}\left(\overline{\rho\, u''^3} + \overline{\rho\, u''v''^2}\right)
\end{aligned} \tag{21-331}$$

Recalling the relations

$$c_v\,\bar{\rho}\tilde{T} + \frac{1}{2}\bar{\rho}\,(\tilde{u}^2+\tilde{v}^2) + \frac{1}{2}\left(\overline{\rho\,''^2} + \overline{\rho\, u''v''^2}\right) = \bar{\rho}\,\tilde{e}_t$$

and

$$c_p\,\bar{\rho}\,\tilde{T} = c_v\,\bar{\rho}\,\tilde{T} + \bar{p}$$

where

$$\bar{p} = \bar{\rho}\, R\,\tilde{T}$$

Then (21-331) is written as

$$\begin{aligned}
\overline{(\rho\, e_t + p)\, u} &= (\bar{\rho}\,\tilde{e}_t + \bar{p})\,\tilde{u} + c_p\,\overline{\rho\, u''T''} + \tilde{u}\,\overline{\rho\, u''^2} + \tilde{v}\,\overline{\rho\, u''v''} \\
&\quad + \frac{1}{2}\left(\overline{\rho\, u''^3} + \overline{\rho\, u''v''^2}\right)
\end{aligned}$$

Similarly, the third term is modified to yield

$$\overline{(\rho\, e_t + p) v} \;=\; (\bar{\rho}\, \tilde{e}_t + \bar{p})\, \tilde{v} + c_p \overline{\rho\, v'' T''} + \tilde{u}\, \overline{\rho\, u'' v''} + \tilde{v}\, \overline{\rho v''^2} + \frac{1}{2}\left(\overline{\rho u''^2 v''} + \overline{\rho\, v''^3} \right)$$

The terms on the right-hand side of the energy equation include shear stress terms such as $u\, \tau_{xx}$. These terms, for example, are written as

$$\overline{u\, \tau_{xx}} = \overline{(\tilde{u} + u'')\, \tau_{xx}} = \tilde{u}\, \bar{\tau}_{xx} + \overline{u'' \tau_{xx}}$$

Finally, the Favre-averaged energy equation is written as

$$\frac{\partial}{\partial t}(\bar{\rho}\, \tilde{e}_t) + \frac{\partial}{\partial x}\Big[(\bar{\rho}\, \tilde{e}_t + \bar{p})\, \tilde{u} + c_p \overline{\rho\, u'' T''} + \tilde{u}\, \overline{\rho u''^2} + \tilde{v}\, \overline{\rho\, u'' v''} + \frac{1}{2}\left(\overline{\rho\, u''^3} + \overline{\rho\, u'' v''^2} \right) \Big] + \frac{\partial}{\partial y}\Big[(\bar{\rho}\, \tilde{e}_t + \bar{p})\, \tilde{v} + c_p \overline{\rho\, v'' T''} + \tilde{u}\, \overline{\rho\, u'' v''} + \tilde{v}\, \overline{\rho\, v''^2} + \frac{1}{2}\left(\overline{\rho\, u''^2 v''} + \overline{\rho\, v''^3} \right) \Big] = \frac{\partial}{\partial x}\Big[\tilde{u}\, \bar{\tau}_{xx} + \overline{u'' \tau_{xx}} + \tilde{v}\, \bar{\tau}_{xy} + \overline{v'' \tau_{xy}} - q_x \Big] + \frac{\partial}{\partial y}\Big[\tilde{u}\, \bar{\tau}_{xy} + \overline{u'' \tau_{xy}} + \tilde{v}\, \bar{\tau}_{yy} + \overline{v'' \tau_{yy}} - \bar{q}_y \Big]$$

and rearranged as

$$\frac{\partial}{\partial t}(\bar{\rho}\, \tilde{e}_t) + \frac{\partial}{\partial x}\left[(\bar{\rho}\, \tilde{e}_t + \bar{p})\, \tilde{u} \right] + \frac{\partial}{\partial y}\left[(\bar{\rho}\, \tilde{e}_t + \bar{p})\, \tilde{v} \right] =$$

$$\frac{\partial}{\partial x}\left[-\tilde{u}\, \overline{\rho\, u''^2} - \tilde{v}\, \overline{\rho\, u'' v''} + \tilde{u}\, \bar{\tau}_{xx} + \tilde{v}\, \bar{\tau}_{xy} \right]$$

$$+ \frac{\partial}{\partial y}\left[-\tilde{u}\, \overline{\rho\, u'' v''} - \tilde{v}\, \overline{\rho\, v''^2} + \tilde{u}\, \bar{\tau}_{xy} + \tilde{v}\, \bar{\tau}_{yy} \right]$$

$$+ \frac{\partial}{\partial x}\left[-c_p \overline{\rho\, u'' T''} - \bar{q}_x \right] + \frac{\partial}{\partial y}\left[-c_p \overline{\rho\, v'' T''} - \bar{q}_y \right]$$

$$+ \frac{\partial}{\partial x}\left[-\frac{1}{2}\left(\overline{\rho\, u''^3} + \overline{\rho\, u'' v''^2} \right) + \overline{u'' \tau_{xx}} + \overline{v'' \tau_{xy}} \right]$$

$$+ \frac{\partial}{\partial y}\left[-\frac{1}{2}\left(\overline{\rho\, u''^2 v''} + \overline{\rho\, v''^3} \right) + \overline{u'' \tau_{xy}} + \overline{v'' \tau_{yy}} \right]$$

The terms involving triple products such as $\overline{\rho\, u''^3}$, $\overline{\rho\, u'' v''^2}$, etc. are typically smaller than other terms and are usually dropped. Therefore, the energy equation is reduced to

$$\frac{\partial}{\partial t}(\rho\,\tilde{e}_t) + \frac{\partial}{\partial x}\left[(\bar{\rho}\,\tilde{e}_t + \bar{p})\,\tilde{u}\right] + \frac{\partial}{\partial y}\left[(\bar{\rho}\,\tilde{e}_t + \bar{p})\,\tilde{v}\right] =$$

$$\frac{\partial}{\partial x}\left[\tilde{u}\left(-\overline{\rho\, u''^2} + \bar{\tau}_{xx}\right) + \tilde{v}\left(-\overline{\rho\, u'' v''} + \bar{\tau}_{xy}\right)\right]$$

$$+\frac{\partial}{\partial y}\left[\tilde{u}\left(-\overline{\rho\, u'' v''} + \bar{\tau}_{xy}\right) + \tilde{v}\left(-\overline{\rho\, v''^2} + \bar{\tau}_{yy}\right)\right]$$

$$+\frac{\partial}{\partial x}\left[-c_p\,\overline{\rho\, u'' T''} - \bar{q}_x\right] + \frac{\partial}{\partial y}\left[-c_p\,\overline{\rho\, v'' T''} - \bar{q}_y\right] \qquad \text{(21-332)}$$

where

$$\bar{q}_x = -\tilde{k}\,\frac{\partial \tilde{T}}{\partial x} = -\frac{c_p}{Pr}\,\tilde{\mu}\,\frac{\partial \tilde{T}}{\partial x}$$

and

$$\bar{q}_y = -\tilde{k}\,\frac{\partial \tilde{T}}{\partial y} = -\frac{c_p}{Pr}\,\tilde{\mu}\,\frac{\partial \tilde{T}}{\partial y}$$

Now the system of equations, composed of the Favre-averaged Navier-Stokes equations, are written in a vector form as

$$\frac{\partial Q}{\partial t} + \frac{\partial E}{\partial x} + \frac{\partial F}{\partial y} = \frac{\partial E_v}{\partial x} + \frac{\partial F_v}{\partial y} \qquad \text{(21-333)}$$

where

$$Q = \begin{bmatrix} \bar{\rho} \\ \bar{\rho}\,\tilde{u} \\ \bar{\rho}\,\tilde{v} \\ \bar{\rho}\,\tilde{e}_t \end{bmatrix} \qquad \text{(21-334)} \qquad E = \begin{bmatrix} \bar{\rho}\,\tilde{u} \\ \bar{\rho}\tilde{u}^2 + \bar{p} \\ \bar{\rho}\,\tilde{u}\tilde{v} \\ (\bar{\rho}\,\tilde{e}_t + \bar{p})\tilde{u} \end{bmatrix} \qquad \text{(21-335)}$$

$$F = \begin{bmatrix} \bar{\rho}\,\tilde{v} \\ \bar{\rho}\,\tilde{v}\,\tilde{u} \\ \bar{\rho}\,\tilde{v}^2 + \bar{p} \\ (\bar{\rho}\,\tilde{e}_t + \bar{p})\tilde{v} \end{bmatrix} \qquad \text{(21-336)}$$

$$E_v = \begin{bmatrix} 0 \\ \bar{\tau}_{xx} - \overline{\rho\, u''^2} \\ \bar{\tau}_{xy} - \overline{\rho\, u'' v''} \\ \tilde{u}\left(\bar{\tau}_{xx} - \overline{\rho\, u''^2}\right) + \tilde{v}\left(\bar{\tau}_{xy} - \overline{\rho\, u'' v''}\right) - \bar{q}_x - c_p \overline{\rho\, u'' T''} \end{bmatrix} \tag{21-337}$$

$$F_v = \begin{bmatrix} 0 \\ \bar{\tau}_{xy} - \overline{\rho\, u'' v''} \\ \bar{\tau}_{yy} - \overline{\rho\, v''^2} \\ \tilde{u}\left(\bar{\tau}_{xy} - \overline{\rho\, u'' v''}\right) + \tilde{v}\left(\bar{\tau}_{yy} - \overline{\rho\, v''^2}\right) - \bar{q}_y - c_p \overline{\rho\, v'' T''} \end{bmatrix} \tag{21-338}$$

Similar to the Reynolds-averaged equations, the Favre-fluctuating terms are related to the Favre-averaged mean quantitites through the introduction of turbulent viscosity. The procedure and definition of terms are provided in the following section.

21.8.1 Turbulent Viscosity

Following the Boussinesq approximation introduced in Section 21.3.1.1, the turbulent shear stresses are related to the rate of strain in a similar way as the laminar shear stresses. Thus, a turbulent (or eddy) viscosity is introduced in the expressions for turbulent shear stresses as follows.

$$-\overline{\rho\, u'' v''} = \mu_t \left(\frac{\partial \tilde{u}}{\partial y} + \frac{\partial \tilde{v}}{\partial x}\right) \tag{21-339}$$

$$-\overline{\rho\, u''^2} = \mu_t \left(2\frac{\partial \tilde{u}}{\partial x} - \frac{2}{3}\nabla\cdot\vec{\tilde{V}}\right) - \frac{2}{3}\bar{\rho}\, k \tag{21-340}$$

$$-\overline{\rho\, v''^2} = \mu_t \left(2\frac{\partial \tilde{v}}{\partial y} - \frac{2}{3}\nabla\cdot\vec{\tilde{V}}\right) - \frac{2}{3}\bar{\rho}\, k \tag{21-341}$$

The stress terms defined above can be expressed in a tensor notation given by

$$-\overline{\rho\, u_i'' u_j''} = \mu_t \left(\frac{\partial \tilde{u}_i}{\partial x_j} + \frac{\partial \tilde{u}_j}{\partial x_i} - \frac{2}{3}\frac{\partial \tilde{u}_k}{\partial x_k}\delta_{ij}\right) - \frac{2}{3}\bar{\rho}\, k\, \delta_{ij} \tag{21-342}$$

Relation (21-342) can be written in the following form as well

$$-\overline{\rho\, u_i'' u_j''} = 2\mu_t \left(\tilde{S}_{ij} - \frac{1}{3}\frac{\partial \tilde{u}_k}{\partial x_k}\delta_{ij}\right) - \frac{2}{3}\bar{\rho}\, k\, \delta_{ij} \tag{21-343}$$

where

$$\tilde{S}_{ij} = \frac{1}{2}\left(\frac{\partial \tilde{u}_i}{\partial x_j} + \frac{\partial \tilde{u}_j}{\partial x_i}\right) \tag{21-344}$$

Turbulent heat flux terms are also written in a form such as to resemble the laminar heat flux term. For this purpose, a turbulent thermal conductivity is introduced such that

$$c_p \overline{\rho\, u''T''} = -k_t \frac{\partial \tilde{T}}{\partial x} \tag{21-345}$$

and

$$c_p \overline{\rho\, v''T''} = -k_t \frac{\partial \tilde{T}}{\partial y} \tag{21-346}$$

It is a common practice to write the turbulent thermal conductivity in terms of turbulent viscosity by the introduction of a turbulent Prandtl number defined as

$$Pr_t = \frac{c_p\, \mu_t}{k_t}$$

This is the same approach taken in the Reynolds-averaged formulation. Typically the value of the turbulent Prandtl number is selected as a constant, and for air is specified as either 0.9 for wall bounded flows or 0.5 for free shear layers.

Now the stress and heat flux terms in (21-337) and (21-338) are redefined as follows.

$$\tilde{\tau}_{xx} = \bar{\tau}_{xx} - \overline{\rho\, u''^2} = (\tilde{\mu} + \mu_t)\left[2\frac{\partial \tilde{u}}{\partial x} - \frac{2}{3}\nabla\cdot\vec{\tilde{V}}\right] - \frac{2}{3}\bar{\rho}\, k \tag{21-347}$$

$$\tilde{\tau}_{xy} = \bar{\tau}_{xy} - \overline{\rho\, u''v''} = (\tilde{\mu} + \mu_t)\left(\frac{\partial \tilde{u}}{\partial y} + \frac{\partial \tilde{v}}{\partial x}\right) \tag{21-348}$$

$$\tilde{\tau}_{yy} = \bar{\tau}_{yy} - \overline{\rho\, v''^2} = (\tilde{\mu} + \mu_t)\left[2\frac{\partial \tilde{v}}{\partial y} - \frac{2}{3}\nabla\cdot\vec{\tilde{V}}\right] - \frac{2}{3}\bar{\rho}\, k \tag{21-349}$$

$$-\tilde{q}_x = -\bar{q}_x - c_p\overline{\rho\, u''T''} = (\tilde{k} + k_t)\frac{\partial \tilde{T}}{\partial x} = c_p\left(\frac{\tilde{\mu}}{Pr} + \frac{\mu_t}{Pr_t}\right)\frac{\partial \tilde{T}}{\partial x} \tag{21-350}$$

$$-\tilde{q}_y = -\bar{q}_y - c_p\overline{\rho\, v''T''} = (\tilde{k} + k_t)\frac{\partial \tilde{T}}{\partial y} = c_p\left(\frac{\tilde{\mu}}{Pr} + \frac{\mu_t}{Pr_t}\right)\frac{\partial \tilde{T}}{\partial y} \tag{21-351}$$

Finally, the FANS equations are written as

$$\frac{\partial Q}{\partial t} + \frac{\partial E}{\partial x} + \frac{\partial F}{\partial y} = \frac{\partial E_v}{\partial x} + \frac{\partial F_v}{\partial y} \tag{21-352}$$

where

$$Q = \begin{bmatrix} \bar{\rho} \\ \bar{\rho}\,\tilde{u} \\ \bar{\rho}\,\tilde{v} \\ \bar{\rho}\,\tilde{e}_t \end{bmatrix} \qquad (21\text{-}353) \qquad E = \begin{bmatrix} \bar{\rho}\,\tilde{u} \\ \bar{\rho}\,\tilde{u}^2 + \bar{p} \\ \bar{\rho}\,\tilde{u}\,\tilde{v} \\ (\bar{\rho}\,\tilde{e}_t + \bar{p})\,\tilde{u} \end{bmatrix} \qquad (21\text{-}354)$$

$$F = \begin{bmatrix} \bar{\rho}\,\tilde{v} \\ \bar{\rho}\,\tilde{v}\,\tilde{u} \\ \bar{\rho}\,\tilde{v}^2 + \bar{p} \\ (\bar{\rho}\,\tilde{e}_t + \bar{p})\,\tilde{v} \end{bmatrix} \qquad (21\text{-}355)$$

$$E_v = \begin{bmatrix} 0 \\ \tilde{\tau}_{xx} \\ \tilde{\tau}_{xy} \\ \tilde{u}\,\tilde{\tau}_{xx} + \tilde{v}\,\tilde{\tau}_{xy} - \tilde{q}_x \end{bmatrix} \qquad (21\text{-}356)$$

$$F_v = \begin{bmatrix} 0 \\ \tilde{\tau}_{yx} \\ \tilde{\tau}_{yy} \\ \tilde{u}\,\tilde{\tau}_{yx} + \tilde{v}\,\tilde{\tau}_{yy} - \tilde{q}_y \end{bmatrix} \qquad (21\text{-}357)$$

Observe that the flux vector formulation given by (21-352) and the flux vectors E, F, E_v, and F_v given by relations (21-353) through (21-357) are in a similar form as the corresponding terms in the Reynolds-averaged formulation, and, indeed, in the same form as the corresponding laminar formulation given by (11-192) and flux vectors (11-193a) through (11-193c), (11-193e), and (11-193f).

21.9 Concluding Remarks

Turbulence is a very complex physical phenomenon. Direct simulation of the governing equations with small scale turbulence for a general configuration at high Reynolds numbers is not currently possible. The available approach for practical applications, at the present, uses various types of turbulence models which relate the average effect of turbulence to mean flow. A tremendous amount of work is still required in turbulence modeling. Currently, a universally accepted turbulence model which can be used in diverse flow regimes does not exist. Some models may perform reasonably well in certain flow regimes but result in inaccurate predictions under different situations. Therefore, turbulence models must be carefully investigated for their range of applicability before being utilized.

An attempt has been made in this chapter to introduce some fundamental concepts of turbulent flow, turbulence models, and limited number of solution schemes. It is hoped that this attempt will be beneficial as a first step toward the inclusion of turbulence into the governing equations of fluid motion. An in-depth review of turbulence may be found in various texts such as [21-25] through [21-27].

Chapter 22

Compact Finite Difference Formulations

22.1 Introductory Remarks

In order to numerically solve a partial differential equation, the differentials appearing in the equation must be approximated by algebraic expressions. Taylor series expansions are typically used to develop these expressions as illustrated in Chapter 2. Several formulations expressed as forward, backward, or central difference approximations with various orders of accuracy can be easily developed. Typically, the expressions for the partial derivatives are developed and expressed solely based on the dependent variable. For example, a second-order and a fourth-order central difference expression for the first derivative are given respectively by

$$\frac{\partial f}{\partial x} = f_i' = \frac{f_{i+1} - f_{i-1}}{2\Delta x} + O(\Delta x)^2 \tag{22-1}$$

and

$$\left.\frac{\partial f}{\partial x}\right|_i = f_i' = \frac{f_{i-2} - 8f_{i-1} + 8f_{i+1} - f_{i+2}}{12\Delta x} + O(\Delta x)^4 \tag{22-2}$$

The grid points involved in the computations are shown in Figure 22.1.

Essentially what is being accomplished in the estimation of the derivative f' at point i is a curve fit through several points of data and subsequently determination of the derivative. In fact, the finite difference approximations can be directly obtained by a curve fit of polynomials as shown in Chapter 2.

Next logical approach to increase the accuracy of the estimates of the derivatives, in particular for problems involving shorter length scales or equivalently high frequencies is to include the influence of the derivative's of neighboring points in the calculations. However, keep in mind that these "neighboring" derivatives are unknowns, and therefore the developed expression will involve more than one unknown. Nonetheless, when this equation is applied to all the points within the domain sufficient equations are produced which can be solved simultaneously for the unknowns. This approach is analogous to the solution of a pde by an implicit scheme (more than one unknown per equation), compared to an explicit scheme (one unknown per equation). The resulting approximation is referred to as *Compact Finite Difference* (CTFD) *formulation.* They have been also referred to as *Hermitian formulations.* The derivation of the CTFD formulation also involves Taylor series expansions. A simple example is used here to illustrate the procedure, and subsequently a general formulation is provided.

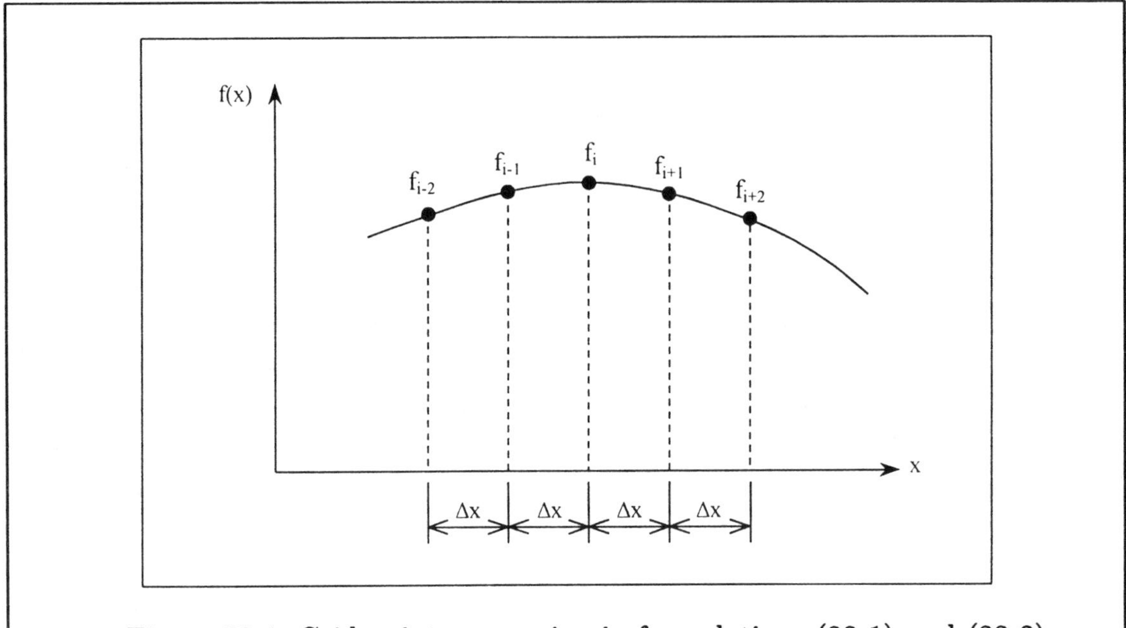

Figure 22-1. Grid points appearing in formulations (22-1) and (22-2).

22.2 Compact Finite Difference Formulations for the First-Order Derivatives

Consider a three-point Hermitian formula involving the first derivative given by

$$H_i = \sum_{m=-1}^{m=+1} (a_m f_{i+m} + b_m f'_{i+m}) = 0 \tag{22-3}$$

or

$$a_1 f_{i+1} + a_o f_i + a_{-1} f_{i-1} + b_1 f'_{i+1} + b_0 f'_i + b_{-1} f'_{i-1} = 0 \tag{22-4}$$

Substition of Taylor series expansions into Equation (22-4) yields

$$a_1 \left[f_i + \Delta x f'_i + \frac{1}{2}(\Delta x)^2 f''_i + \frac{1}{6}(\Delta x)^3 f'''_i + \frac{1}{24}(\Delta x)^4 f^{iv}_i + \frac{1}{120}(\Delta x)^5 f^{v}_i + \right.$$

$$\frac{1}{720}(\Delta x)^6 f_i^{vi} + O(\Delta x)^7\Big] + a_0 f_i + a_{-1}\Big[f_i - \Delta x f_i' + \frac{1}{2}(\Delta x)^2 f_i'' - \frac{1}{6}(\Delta x)^3 f_i''' +$$

$$\frac{1}{24}(\Delta x)^4 f_i^{iv} - \frac{1}{120}(\Delta x)^5 f_i^{v} + \frac{1}{720}(\Delta x)^6 f_i^{vi} + O(\Delta x)^7\Big] + b_1\Big[f_i' + \Delta x f_i'' +$$

$$\frac{1}{2}(\Delta x)^2 f_i''' + \frac{1}{6}(\Delta x)^3 f_i^{iv} + \frac{1}{24}(\Delta x)^4 f_i^{v} + \frac{1}{120}(\Delta x)^5 f_i^{vi} + O(\Delta x)^6\Big] + b_o f_i' +$$

$$b_{-1}\Big[f_i' - \Delta x f_i'' + \frac{1}{2}(\Delta x)^2 f_i''' - \frac{1}{6}(\Delta x)^3 f_i^{iv} + \frac{1}{24}(\Delta x)^4 f_i^{v} -$$

$$\frac{1}{120}(\Delta x)^5 f_i^{vi} + O(\Delta x)^6\Big] = 0 \tag{22-5}$$

or

$$(a_1 + a_0 + a_{-1})f_i + [(a_1 - a_{-1})\Delta x + b_1 + b_0 + b_{-1}]\, f_i' + \Big[\frac{1}{2}(\Delta x)^2(a_1 + a_{-1}) +$$

$$\Delta x(b_1 - b_{-1})\Big] f_i'' + \Big[\frac{1}{6}(\Delta x)^3(a_1 - a_{-1}) + \frac{1}{2}(\Delta x)^2(b_1 + b_{-1})\Big] f_i''' +$$

$$\Big[\frac{1}{24}(\Delta x)^4(a_1 + a_{-1}) + \frac{1}{6}(\Delta x)^3(b_1 - b_{-1})\Big] f_i^{iv} + \Big[\frac{1}{120}(\Delta x)^5(a_1 - a_{-1}) +$$

$$\frac{1}{24}(\Delta x)^4(b_1 + b_{-1})\Big] f_i^{v} + \Big[\frac{1}{720}(\Delta x)^6(a_1 + a_{-1}) +$$

$$\frac{1}{120}(\Delta x)^5(b_1 - b_{-1})\Big] f_i^{vi} = 0 \tag{22-6}$$

Now if the relation given by Equation (22-6) is exact, then the coefficients in (22-6) should be zero. In reality, however, only a few of the coefficients are set to zero, and the remaining higher order terms are truncated and will form the truncation error (TE). To obtain a third-order scheme, set the coefficients of f_i, f_i', f_i'', and f_i''' equal to zero.

$$a_1 + a_0 + a_{-1} = 0 \tag{22-7}$$

$$\Delta x(a_1 - a_{-1}) + b_1 + b_0 + b_{-1} = 0 \tag{22-8}$$

$$\frac{1}{2}(\Delta x)^2(a_1 + a_{-1}) + (\Delta x)(b_1 - b_{-1}) = 0 \tag{22-9}$$

$$\frac{1}{6}(\Delta x)^3(a_1 - a_{-1}) + \frac{1}{2}(\Delta x)(b_1 + b_{-1}) = 0 \tag{22-10}$$

The coefficients are a_1, a_0, a_{-1}, and b_0 are solved in terms of b_1, and b_{-1} to yield

$$a_1 = \frac{1}{2\Delta x}(-5b_1 - b_{-1}) \tag{22-11}$$

$$a_0 = \frac{2}{\Delta x}(b_1 - b_{-1}) \tag{22-12}$$

$$a_{-1} = \frac{1}{2\Delta x}(b_1 + 5b_{-1}) \tag{22-13}$$

$$b_o = 2(b_1 + b_{-1}) \tag{22-14}$$

The truncation error from (22-6) is

$$TE = \left[\frac{1}{24}(\Delta x)^4(a_1 + a_{-1}) + \frac{1}{6}(\Delta x)^3(b_1 - b_{-1})\right] f_i^{iv} + \left[\frac{1}{120}(\Delta x)^5(a_1 - a_{-1}) + \frac{1}{24}(\Delta x)^4(b_1 + b_{-1})\right] f_i^{v} + \left[\frac{1}{720}(\Delta x)^6(a_1 + a_{-1}) + \frac{1}{120}(\Delta x)^5(b_1 - b_{-1})\right] f_i^{vi} + \ldots.$$

or with the substitution of (22-11) through (22-14)

$$TE = \frac{1}{12}(\Delta x)^3(b_1 - b_{-1})\, f_i^{iv} + \frac{1}{60}(\Delta x)^4(b_1 + b_{-1})\, f_i^{v} + \frac{1}{180}(\Delta x)^5(b_1 - b_{-1})\, f_i^{vi} + \ldots \tag{22-15}$$

The formulation (22-4) can be written as

$$\frac{a_1}{b_1} f_{i+1} + \frac{a_0}{b_1} f_i + \frac{a_{-1}}{b_1} f_{i-1} + f'_{i+1} + \frac{b_o}{b_1} f'_i + \frac{b_{-1}}{b_1} f'_{i-1} = 0 \tag{22-16}$$

where the equation is divided by b_1. Now, the expressions given by (22-11) through (22-14) are redefined as

$$\frac{a_1}{b_1} = \frac{1}{2\Delta x}(-5 - \alpha) \tag{22-17}$$

$$\frac{a_0}{b_1} = \frac{2}{\Delta x}(1 - \alpha) \tag{22-18}$$

$$\frac{a_{-1}}{b_1} = \frac{1}{2\Delta x}(1 + 5\alpha) \tag{22-19}$$

$$\frac{b_o}{b_1} = 2(1 + \alpha) \tag{22-20}$$

where $\alpha = b_{-1}/b_1$.

Substitutions of (22-17) through (22-20) into (22-16) yield an equation for the derivatives f'_{i-1}, f'_i, and f'_{i+1}, as follows

$$\alpha f'_{i-1} + 2(1+\alpha) f'_i + f'_{i+1} = -\frac{1}{2\Delta x}(1+5\alpha) f_{i-1} - \frac{2}{\Delta x}(1-\alpha) f_i + \frac{1}{2\Delta x}(5+\alpha) f_{i+1} + TE \tag{22-21}$$

where

$$TE = \frac{1}{12}(\Delta x)^3(1-\alpha) f_i^{iv} + \frac{1}{60}(\Delta x)^4(1+\alpha) f_i^{v} + \frac{1}{180}(\Delta x)^5(1-\alpha) f_i^{vi} + .. \tag{22-22}$$

Note that for $\alpha = 1$ the leading error coefficient vanishes and the scheme becomes fourth-order, whereas, for other selections of α's the scheme is third-order. For $\alpha = 1$ the formulation is

$$f'_{i-1} + 4f'_i + f'_{i+1} = -\frac{3}{\Delta x}(f_{i-1} - f_{i+1}) \tag{22-23}$$

Now a general five-point formulation for the approximation of the first derivative can be written as

$$\beta f'_{i-2} + \alpha f'_{i-1} + f'_i + \alpha f'_{i+1} + \beta f'_{i+2} = a\frac{f_{i+1} - f_{i-1}}{2\Delta x} + b\frac{f_{i+2} - f_{i-2}}{4\Delta x} + c\frac{f_{i+3} - f_{i-3}}{6\Delta x} \tag{22-24}$$

The relations between the coefficients in Equation (22-24) are established by matching of the coefficients obtained by the substitution of the Taylor series expansions. The procedure is illustrated for the second-, fourth-, and sixth-order schemes. By retaining higher order terms in the Taylor series, the relations for the eighth- and tenth-order schemes can be established.

The Taylor series expansion of the terms on the left-hand side of (22-24) is

$$f'_{i-2} = f'_i - 2\Delta x f''_i + \frac{4(\Delta x)^2}{2!} f'''_i - \frac{8(\Delta x)^3}{3!} f_i^{iv} + \frac{16(\Delta x)^4}{4!} f_i^{v} \tag{22-25}$$

$$f'_{i-1} = f'_i - \Delta x f''_i + \frac{(\Delta x)^2}{2!} f'''_i - \frac{(\Delta x)^3}{3!} f_i^{iv} + \frac{(\Delta x)^4}{4!} f_i^{v} \tag{22-26}$$

$$f'_{i+1} = f'_i + \Delta x f''_i + \frac{(\Delta x)^2}{2!} f'''_i + \frac{(\Delta x)^3}{3!} f_i^{iv} + \frac{(\Delta x)^4}{4!} f_i^{v} \tag{22-27}$$

$$f'_{i+2} = f'_i + 2\Delta x f''_i + \frac{4(\Delta x)^2}{2!} f''_i + \frac{8(\Delta x)^3}{3!} f_i^{iv} + \frac{16(\Delta x)^4}{4!} f_i^{v} \tag{22-28}$$

The left-hand side of (22-24) is

$$\beta f'_{i-2} + \alpha f'_{i-1} + f'_i + \alpha f'_{i+1} + \beta f'_{i+2} = \beta(f'_{i-2} + f'_{i+2}) + \alpha(f'_{i-1} + f'_{i+1}) + f'_i \tag{22-29}$$

Substitution of expansions (22-25) through (22-28) into (22-29) yields

$$\beta\left[2f_i' + 4(\Delta x)^2 f_i''' + \frac{32(\Delta x)^4}{4!} f_i^v\right] + \alpha\left[2f_i' + (\Delta x)^2 f_i''' + \frac{2(\Delta x)^4}{4!} f_i^v\right] + f_i' =$$

$$(2\alpha + 2\beta + 1) f_i' + (\Delta x)^2(\alpha + 4\beta) f_i''' + \frac{2(\Delta x)^4}{4!}(\alpha + 16\beta) f_i^v \qquad (22\text{-}30)$$

The required Taylor series expansions for the terms on the right-hand side of (22-24) are

$$f_{i-3} = f_i - 3\Delta x f_i' + \frac{9(\Delta x)^2}{2!} f_i'' - \frac{27(\Delta x)^3}{3!} f_i''' + \frac{81(\Delta x)^4}{4!} f_i^{iv} - \frac{243(\Delta x)^5}{5!} f_i^v \quad (22\text{-}31)$$

$$f_{i-2} = f_i - 2\Delta x f_i' + \frac{4(\Delta x)^2}{2!} f_i'' - \frac{8(\Delta x)^3}{3!} f_i''' + \frac{16(\Delta x)^4}{4!} f_i^{iv} - \frac{32(\Delta x)^5}{5!} f_i^v \quad (22\text{-}32)$$

$$f_{i-1} = f_i - \Delta x\, f_i' + \frac{(\Delta x)^2}{2!} f_i'' - \frac{(\Delta x)^3}{3!} f_i''' + \frac{(\Delta x)^4}{4!} f_i^{iv} - \frac{(\Delta x)^5}{5!} f_i^v \quad (22\text{-}33)$$

$$f_{i+1} = f_i + \Delta x f_i' + \frac{(\Delta x)^2}{2!} f_i'' + \frac{(\Delta x)^3}{3!} f_i''' + \frac{(\Delta x)^4}{4!} f_i^{iv} + \frac{(\Delta x)^5}{5!} f_i^v \quad (22\text{-}34)$$

$$f_{i+2} = f_i + 2\Delta x f_i' + \frac{4(\Delta x)^2}{2!} f_i'' + \frac{8(\Delta x)^3}{3!} f_i''' + \frac{16(\Delta x)^4}{4!} f_i^{iv} + \frac{32(\Delta x)^5}{5!} f_i^v \quad (22\text{-}35)$$

$$f_{i+3} = f_i + 3\Delta x f_i' + \frac{9(\Delta x)^2}{2!} f_i'' + \frac{27(\Delta x)^3}{3!} f_i''' + \frac{81(\Delta x)^4}{4!} f_i^{iv} + \frac{243(\Delta x)^5}{5!} f_i^v \quad (22\text{-}36)$$

Expansions (22-31) through (22-36) are substituted into the right-hand side of (22-24) to provide

$$\frac{a}{2\Delta x}(f_{i+1} - f_{i-1}) + \frac{b}{4\Delta x}(f_{i+2} - f_{i-2}) + \frac{c}{6\Delta x}(f_{i+3} - f_{i-3}) =$$

$$\frac{a}{2\Delta x}\left[\ 2\Delta x f_i' + 2\frac{(\Delta x)^3}{3!} f_i''' + 2\frac{(\Delta x)^5}{5!} f_i^v\right] + \frac{b}{4\Delta x}\left[4\Delta x f_i' + 2\frac{8(\Delta x)^3}{3!} f_i'''\right.$$

$$\left. + 2\,\frac{32(\Delta x)^5}{5!} f_i^v\right] + \frac{c}{6\Delta x}\left[6\Delta x f_i' + 2\frac{27(\Delta x)^3}{3!} f_i''' + 2\frac{243(\Delta x)^5}{5!} f_i^v\right]$$

$$= \ (a + b + c) f_i' + \frac{(\Delta x)^2}{3!}(a + 4b + 9c) f_i''' + \frac{(\Delta x)^4}{5!}(a + 16b + 81c) f_i^v \quad (22\text{-}37)$$

Now from expressions (22-30) and (22-37) one obtains

$$(2\alpha + 2\beta + 1)\, f_i' + (\Delta x)^2(\alpha + 4\beta) f_i''' + \frac{2(\Delta x)^4}{4!}(\alpha + 16\beta)\, f_i^v =$$

$$(a + b + c) f_i' + \frac{(\Delta x)^2}{3!}(a + 4b + 9c) f_i''' + \frac{(\Delta x)^4}{5!}(a + 16b + 81c) f_i^v \quad (22\text{-}38)$$

Setting the coefficients of the differentials equal, the following constraints between the coefficients are established

$$\text{Second-order:} \quad 1 + 2\alpha + 2\beta = a + b + c \qquad (22\text{-}39)$$

$$\text{Fourth-order:} \quad 2\frac{3!}{2!}(\alpha + 2^2\beta) = a + 2^2 b + 3^2 c \qquad (22\text{-}40)$$

$$\text{Sixth-order:} \quad 2\frac{5!}{4!}(\alpha + 2^4\beta) = a + 2^4 b + 3^4 c \qquad (22\text{-}41)$$

The constraints for higher-order terms are :

$$\text{Eighth-order:} \quad 2\frac{7!}{6!}(\alpha + 2^6\beta) = a + 2^6 b + 3^6 c \qquad (22\text{-}42)$$

$$\text{Tenth-order:} \quad 2\frac{9!}{8!}(\alpha + 2^8\beta) = a + 2^8 b + 3^8 c \qquad (22\text{-}43)$$

To obtain the error term of a particular scheme, the truncated terms are considered. For example, for the fourth-order scheme, the truncated terms are

$$TE = [\frac{1}{5!}(a + 2^4 b + 3^4 c) - \frac{2}{4!}(\alpha + 2^4\beta)]\,(\Delta x)^4 f_i^v \qquad (22\text{-}44)$$

For the values of $\beta = 0$, $a = \frac{2}{3}(\alpha + 2), b = \frac{1}{3}(4\alpha - 1)$, and $c = 0$

$$TE = \frac{4}{5!}(3\alpha - 1)(\Delta x)^4 f_i^v \qquad (22\text{-}45)$$

Depending on how α and β are specified, either a tridiagonal system or a pentadiagonal system will be generated from Equation (22-24). For the special case of $\beta = 0$, the resulting system of equation will be tridiagonal. Several special cases are presented in this section.

For $\beta = 0$ and $c = 0$, a one-parameter (in terms of α) family of fourth-order tridiagonal systems can be produced where $a = \frac{2}{3}(\alpha + 2)$ and $b = \frac{1}{3}(4\alpha - 1)$. As seen previously, the truncation error of the scheme is $TE = \frac{4}{5!}(3\alpha - 1)(\Delta x)^4 f_i^v$

Observe that for $\alpha = \frac{1}{4}$, then b = 0, and $a = \frac{3}{2}$, and one has

$$\frac{1}{4} f_{i-1}' + f_i' + \frac{1}{4} f_{i+1}' = \frac{3}{2}\, \frac{f_{i+1} - f_{i-1}}{2\Delta x} \qquad (22\text{-}46)$$

which is identified to Equation (22-23). Furthermore, note that the selection of $\alpha = 1/4$, $b = 0$, and $a = 3/2$, satisfies the requirement given by relation (22-40).

The one-parameter family of schemes are summarized in Table 22.1

Scheme	Eq.	α	β	a	b	c	Order of accuracy
I	22-47	α	0	$\frac{2}{3}(\alpha+2)$	$\frac{1}{3}(4\alpha-1)$	0	4th
II	22-48	α	0	$\frac{1}{6}(\alpha+9)$	$\frac{1}{15}(32\alpha-9)$	$\frac{1}{10}(-3\alpha+1)$	6th
III	22-49	α	$\frac{1}{20}(-3+8\alpha)$	$\frac{1}{6}(12-7\alpha)$	$\frac{1}{150}(568\alpha-183)$	$\frac{1}{50}(9\alpha-4)$	8th

Scheme	Eq.	Truncation error	Max LHS stencil size	Max RHS stencil size
I	22-47	$\frac{4}{5!}(3\alpha-1)(\Delta x)^4 f_i^v$	3	5
II	22-48	$\frac{12}{7!}(-8\alpha+3)(\Delta x)^6 f_i^{vii}$	3	7
III	22-49	$\frac{144}{9!}(2\alpha-1)(\Delta x)^8 f_i^{ix}$	5	7

Table 22.1. Selected one-parameter family of compact finite difference schemes for the first-order derivative.

The following observations can be made with regard to the schemes which are given in Table 22.1:

1. Scheme I
   (a) For $\alpha = 0$ the values of a and b are $\frac{4}{3}$ and -$\frac{1}{3}$, respectively, and relation (22-24) provides

$$f_i' = \frac{f_{i-2} - 8f_{i-1} + 8f_{i+1} - f_{i+2}}{12\Delta x} \tag{22-50}$$

   which is the fourth-order central difference formulation.

   (b) For $\alpha = \frac{1}{3}$ and the resulting values of $a = \frac{14}{9}$ and $b = \frac{1}{9}$, the scheme becomes formally sixth-order due to vanishing of the leading term in the truncation error.

2. Scheme II
   (a) For $\alpha = \frac{1}{3}$, then $a = \frac{14}{9}$, $b = \frac{1}{9}$, and $c = 0$, and the result is identical to formulation of Ib. Therefore, the formulation of Ib can be considered as a member of II.

   (b) For $\alpha = \frac{3}{8}$, then $a = \frac{75}{48}$, $b = \frac{1}{5}$, and $c = -\frac{1}{80}$, and the scheme is eighth-order.

3. Scheme III.
   (a) For $\alpha = \frac{3}{8}$, $\beta = 0$, $a = \frac{75}{48}$, $b = \frac{1}{5}$, $c = -\frac{1}{80}$, the scheme is identical to IIb.

   (b) For $\alpha = \frac{1}{2}$, $\beta = \frac{1}{20}$, $a = \frac{17}{12}$, $b = \frac{101}{150}$, and $c = \frac{1}{100}$, the scheme becomes tenth-order accurate.

22.2.1 One-Sided Approximations

Several schemes identified for the first derivative based on the central difference approximation of (22-24) will encounter difficulties for nonperiodic boundary conditions. Typically, one-sided approximations are used at the boundaries. Considering points $i = 1$ and $i = \text{IM}$ to be the boundary points, the following general expressions for the first derivative may be written.

$$f_1' + \alpha f_2' = \frac{1}{\Delta x}(af_1 + bf_2 + cf_3 + df_4) \tag{22-51}$$

and

$$f_{IM}' + \alpha f_{IM1}' = \frac{1}{\Delta x}(af_{IM} + bf_{IMM1} + cf_{IMM2} + df_{IMM3}) \tag{22-52}$$

The relation between the coefficients are established in the same manner as presented previously and are given by the following.

$$\text{Second-order:} \quad \begin{aligned} a &= -\frac{1}{2}(3 + \alpha + 2d) \\ b &= 2 + 3d \\ c &= -\frac{1}{2}(1 - \alpha + 6d) \end{aligned}$$

and the truncation error is

$$TE = \frac{1}{3!}(2 - \alpha - 6d)(\Delta x)^2 f_1'' \tag{22-53}$$

$$\text{Third-order:} \quad \begin{aligned} a &= -\frac{1}{6}(11 + 2\alpha) \\ b &= \frac{1}{2}(6 - \alpha) \\ c &= \frac{1}{2}(2\alpha - 3) \\ d &= \frac{1}{6}(2 - \alpha) \end{aligned}$$

with the truncation error of

$$TE = \frac{2}{4!}(\alpha - 3)(\Delta x)^3 f_1^{iv} \tag{22-54}$$

$$\text{Fourth-order:} \quad \begin{aligned} \alpha &= 3 \\ a &= -\frac{17}{6} \end{aligned}$$

$$b = \frac{3}{2}$$

$$c = \frac{3}{2}$$

$$d = -\frac{1}{6}$$

and

$$TE = \frac{6}{5!}(\Delta x)^4 f_1^v \tag{22-55}$$

22.3 Compact Finite Difference Formulations for the Second-Order Derivatives

A similar expression to that of (22-24) can be written for the second derivative as follows

$$\beta f_{i-2}'' + \alpha f_{i-1}'' + f_i'' + \alpha f_{i+1}'' + \beta f_{i+2}'' = a\frac{f_{i+1} - 2f_i + f_{i-1}}{(\Delta x)^2} + b\frac{f_{i+2} - 2f_i + f_{i-2}}{4(\Delta x)^2} + c\frac{f_{i+3} - 2f_i + f_{i-3}}{9(\Delta x)^2} \tag{22-56}$$

The relations among the coefficients are established as

Second-order: $$2\alpha + 2\beta + 1 = a + b + c \tag{22-57}$$

Fourth-order: $$\frac{4!}{2!}(\alpha + 2^2\beta) = a + 2^2 b + 3^2 c \tag{22-58}$$

Sixth-order: $$\frac{6!}{4!}(\alpha + 2^2\beta) = a + 2^4 b + 3^4 c \tag{22-59}$$

Eighth-order: $$\frac{8!}{6!}(\alpha + 2^6\beta) = a + 2^6 b + 3^6 c \tag{22-60}$$

Tenth-order: $$\frac{10!}{8!}(\alpha + 2^8\beta) = a + 2^8 b + 2^8 c \tag{22-61}$$

Several formulations for the second derivative are provided in Table 22.2.

Scheme	Eq.	α	β	a	b	c	Order of accuracy
I	22-62	α	0	$\frac{4}{3}(1-\alpha)$	$\frac{1}{3}(10\alpha-1)$	0	4th
II	22-63	α	β	$\frac{1}{4}(6-9\alpha-12\beta)$	$\frac{1}{5}(-3+24\alpha-6\beta)$	$\frac{1}{20}(2-11\alpha+124\beta)$	6th
III	22-64	α	$\frac{1}{214}(38\alpha-9)$	$\frac{1}{428}(696-1191\alpha)$	$\frac{1}{535}(245\alpha-294)$	$\frac{1}{2140}(1179\alpha-344)$	8th

Scheme	Eq.	Truncation error	Max LHS stencil size	Max RHS stencil size
I	22-62	$\frac{4}{6!}(11\alpha-2)\,(\Delta x)^4 f^{vi}$	3	5
II	22-63	$-\frac{8}{8!}(9-38\alpha+214\beta)\,(\Delta x)^6 f^{viii}$	5	7
III	22-64	$\frac{1}{2696400}(899\alpha-334)\,(\Delta x)^8 f^{x}$	5	7

Table 22.2. Selected compact finite difference schemes for the second order derivative.

Observe that for the scheme given by Equation (22-62), when $\alpha = 0$, the values of a and b are determined to be $\frac{4}{3}$ and $-\frac{1}{3}$ respectively. Subsequently Equation (22-56) provides

$$f_i'' = \frac{-f_{i-2} + 16f_{i-1} - 30f_i + 16f_{i+1} - f_{i+2}}{12(\Delta x)^2} \tag{22-65}$$

which is the fourth-order central difference approximation of f_i''. For the case of $\alpha = \frac{2}{11}$, $a = \frac{12}{11}$, and $b = \frac{3}{11}$, the scheme is formally sixth-order accurate.

22.4 Error Analysis

To investigate the resolution characteristics of finite difference approximations and to perform error analysis, the Fourier analysis approach is undertaken. This procedure is relatively simple and provides an effective method to quantify the resolution issue of the finite difference scheme as well as error analysis.

To further simplify the Fourier analysis of error, a function which is periodic over the domain is assumed. Now, the function is expressed with a complex Fourier series by

$$f(x) = \sum_{k=-IM/2}^{k=IM/2} F_k \, e^{I\frac{2\pi k x}{L}} \tag{22-66}$$

where $I = \sqrt{-1}$, L is the length of the domain, x is the independent variable, and F_k is the kth component of complex Fourier coefficient. Furthermore, $\Delta x = \frac{L}{IM}$ is the spatial grid size, $s = \frac{x}{\Delta x}$ and $K = \frac{2\pi x}{L} = \frac{2\pi k}{IM}$ are defined as the scaled wave number. Now,

$$f(s) = \sum_{k=-IM/2}^{k=IM/2} F_k \, e^{IKs} \tag{22-67}$$

and the first derivative of (22-67) is

$$f'(s) = \sum_{k=-IM/2}^{k=IM/2} F_k IK \, e^{IKs} \tag{22-68}$$

or

$$f'(s) = \sum_{k=-IM/2}^{k=IM/2} F_k' \, e^{IKs} \tag{22-69}$$

where

$$F_k' = I\,K\,F_k \tag{22-70}$$

Now, consider a finite difference approximation of the first derivative of f as

$$f_{fd}'(s) = \sum_{k=-IM/2}^{IM/2} F_{k_{fd}}' e^{IKs} \tag{22-71}$$

or

$$f_{fd}'(s) = \sum_{k=-IM/2}^{IM/2} F_k IK' e^{IKs} \tag{22-72}$$

where

$$F_{k_{fd}}' = I\,K'\,F_k \tag{22-73}$$

and $K' = K'(k)$ is the modified wave number. One obtains the exact value if

$$F_{k_{fd}}' = F_k' \tag{22-74}$$

or

$$K' = K \tag{22-75}$$

which would be a straight line in a wave number, modified wave number coordinate.

Now, define the error as

$$ER = \frac{K - K'}{K} \tag{22-76}$$

from which one may write

$$K' = K(1 - ER) \tag{22-77}$$

From the definition given by (22-76), the following may be written

$$ER = \frac{f'(s) - f_{fd}'(s)}{f'(s)} = \frac{\Sigma F_k IK \, e^{IKs} - \Sigma \, F_k I \, K' e^{IKs}}{\Sigma \, F_k IK \, e^{IKs}} \tag{22-78}$$

Now, consider an example where a sine wave is used. In this case, the Fourier series is reduced to one term only, therefore,

$$f(x) = \sin(Ks) = \sin(K\frac{x}{\Delta x}) \tag{22-79}$$

and

$$f'(x) = \frac{K}{\Delta x}\cos(K\frac{x}{\Delta x}) = \frac{K}{\Delta x}\cos(Ks) \tag{22-80}$$

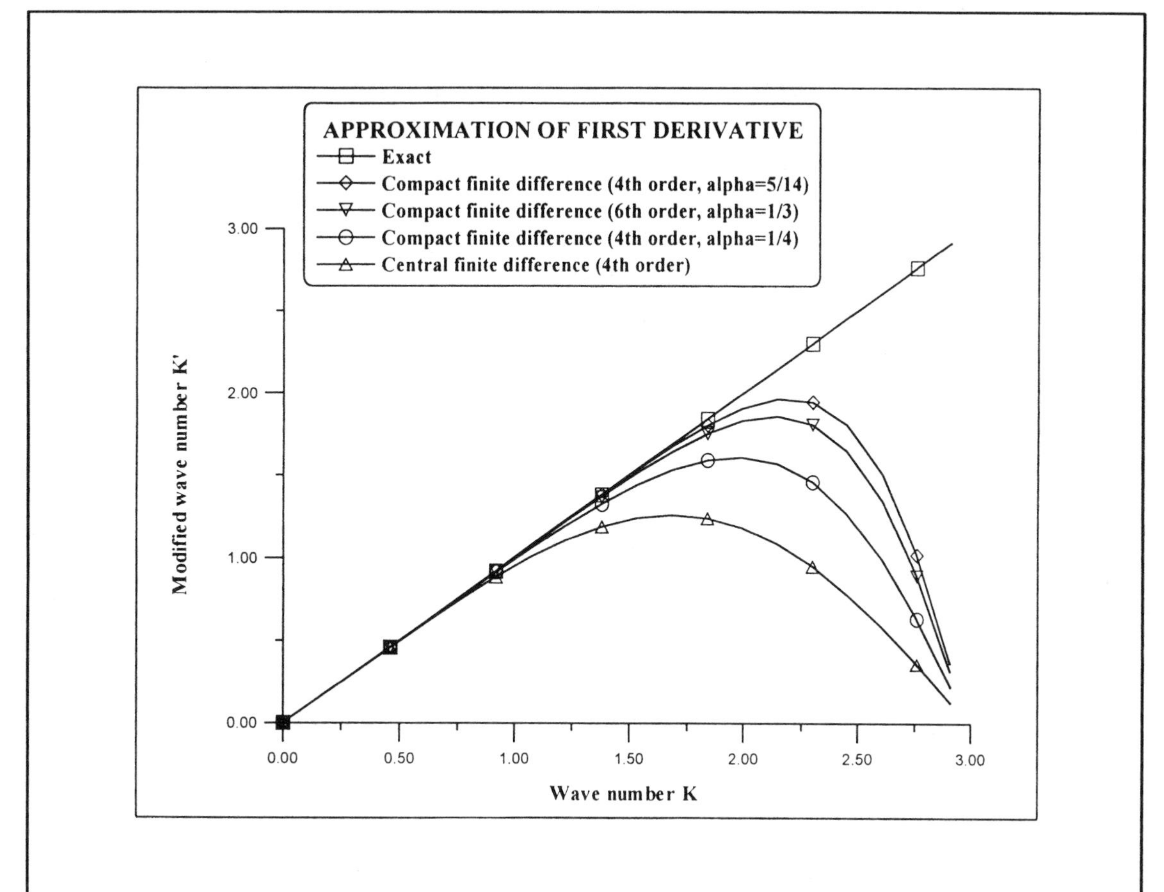

Figure 22-2. Comparison of several schemes for the approximation of the first derivative of (22-79).

A comparison of the wave number K versus the modified wave number K' is shown in Figure 22-2 for several schemes. It is evident that the compact finite difference follows the exact differentiation over a wider range of wave numbers compared to the fourth-order central difference approximation. The approximate value of maximum wave number for well resolved waves (k_r) is provided in Table 22.3. A measure of resolving efficiency of a scheme may be defined by $e = \frac{k_r}{\pi}$. For example, as seen from Table 22.3, the resolving efficiency of the fourth-order scheme with $\alpha = 5/14$ is higher than the resolving efficiency of the sixth-order scheme with $\alpha = \frac{1}{3}$.

Scheme	k_r	e
Central finite difference (fourth-order)	0.6	0.191
Compact finite difference (fourth-order, $\alpha = \frac{1}{4}$)	1.0	0.318
Compact finite difference (sixth-order, $\alpha = \frac{1}{3}$)	1.25	0.398
Compact finite difference (fourth-order, $\alpha = \frac{5}{14}$)	1.50	0.477

Table 22.3. Comparison of the resolving efficiency of several schemes.

22.5 Application to a Hyperbolic Equation

A simple linear hyperbolic equation is used to obtain the solution by several schemes and compare the solutions to each other and the analytical solution. The numerical schemes considered are (1) a first-order upwind scheme, (2) a modified four stage Runge-Kutta scheme where the convective terms are approximated by a fourth-order central difference expression, and (3) a modified four stage Runge-Kutta scheme where the convective terms are approximately by a sixth-order central compact finite difference scheme.

Consider the first-order wave equation

$$\frac{\partial u}{\partial t} = -a\frac{\partial u}{\partial x}, \qquad a > 0 \tag{22-81}$$

where a is specified to be $40\pi\ m/s$, and the domain of solution is $0 \leq x \leq 10\pi\ m$. An initial wave is specified as

$$\begin{aligned} u(x,0) &= 0.0 & 0 \leq x < 2\pi \\ u(x,0) &= 1.0 - \cos x & 2\pi \leq x < 4\pi \\ u(x,0) &= 0.0 & 4\pi \leq x < 10\pi \end{aligned} \tag{22-82}$$

The spatial grid is

$$\Delta x = \frac{10\pi}{IM - 1}$$

where $IM = 51$.

Several temporal steps (or corresponding Courant numbers) are specified as follow.

1. (a) $\Delta t = 0.005$ sec, $c = 1.0$
   (b) $\Delta t = 0.0025$ sec, $c = 0.5$
   (c) $\Delta t = 0.00125$ sec, $c = 0.25$

The explicit first-order upwind scheme applied to Equation (22-81) yields the following FDE.

$$u_i^{n+1} = u_i^n - c\,(u_i^n - u_{i-1}^n) \tag{22-83}$$

The modified Runge-Kutta scheme is given by

$$u_i^{(1)} = u_i^n \tag{22-84}$$

$$u_i^{(2)} = u_i^n - a\frac{\Delta t}{4}(\frac{\partial u}{\partial x})_i^{(1)} \tag{22-85}$$

$$u_i^{(3)} = u_i^n - a\frac{\Delta t}{3}(\frac{\partial u}{\partial x})_i^{(2)} \tag{22-86}$$

$$u_i^{(4)} = u_i^n - a\frac{\Delta t}{2}(\frac{\partial u}{\partial x})_i^{(3)} \tag{22-87}$$

$$u_i^{n+1} = u_i^n - a\Delta t(\frac{\partial u}{\partial x})_i^{(4)} \tag{22-88}$$

Two options are considered in the approximation of the first-order derivative on the right hand side of Equations (22-85) through (22-88). In the first option, a fourth-order central difference approximation given by

$$\frac{\partial u}{\partial x} = \frac{u_{i-2} - 8u_{i-1} + 8u_{i+1} - u_{i+2}}{12\Delta x} + 0(\Delta x)^4 \tag{22-89}$$

is used for $3 \le i \le IM - 2$, where a second-order scheme is implemented at $i = 2$ and $i = IM - 1$.

For the second option, the derivatives are evaluated by the central compact formulation given by Equation (22-48) with $\alpha = 1/3$, $a = 14/9$, and $b = 1/9$, which is formally six-order accurate.

Before the inspection of numerical solutions, a review of several observations with regard to the numerical schemes to be used is in order.

1. Recall that the error term for the explicit first-order upwind scheme is given by Equation (4-77), repeated here for convenience.

$$ER = \frac{a\Delta x}{2}(1-c)\frac{\partial^2 u}{\partial x^2} - \frac{a(\Delta x)^2}{6}(2c^2 - 3c + 1)\frac{\partial^3 u}{\partial x^3} + O[(\Delta x)^3, (\Delta x)(\Delta t)^2, (\Delta x)^2(\Delta t), (\Delta x)^3] \tag{22-90}$$

Note that the dominant term involves a second derivative $\partial^2 u / \partial x^2$, and therefore the error produced in the solution is dissipative. That is, as the Courant number is decreased from its maximum value of one, the amplitude of the wave is decreased and the wave is dissipated to the neighboring points. The solution at Courant number of 1 is exact as seen from Equation (22-90), where the error becomes zero.

2. The dominant error in schemes which utilize central difference approximation for the convective term is dispersive and results in oscillations in the solution in the vicinity of large gradients. These observations are clearly evident in Figures 22-3 and 22-4. The solutions by the first-order upwind scheme for the three different Courant numbers (or time steps) are shown in Figure 22-3. The solution for Courant number of 1 is exact as seen in Figure 22-3a. The dissipation error and its effect on the solution is shown in Figures 22-3b and 22-3c. The solutions by the modified Runge-Kutta scheme with central difference approximation of the convective term are given in Figures 22-4a through 22-4c. Several schemes are available to reduce the oscillations, for example, addition of a fourth-order damping term or incorporation of a TVD model into the formulation as shown in Chapter 6. The solutions by the Runge-Kutta scheme with central compact formulations are illustrated in Figure 22-5. The solutions retain the original wave shape for all Courant numbers specified in the problem. A comparison of numerical solutions with each other and the analytical solution are shown in Figure 22-6.

The solution by the compact formulation is well behaved for all the three specified Courant numbers. Of course, this increase in accuracy is accompanied by an increase in computation time. However, it should be stated that in some applications it will be necessary to implement the compact formulation due to its resolution efficiency of small scales, which may be present in these applications. Example of such applications includes problems in computational acoustics and direct numerical simulation.

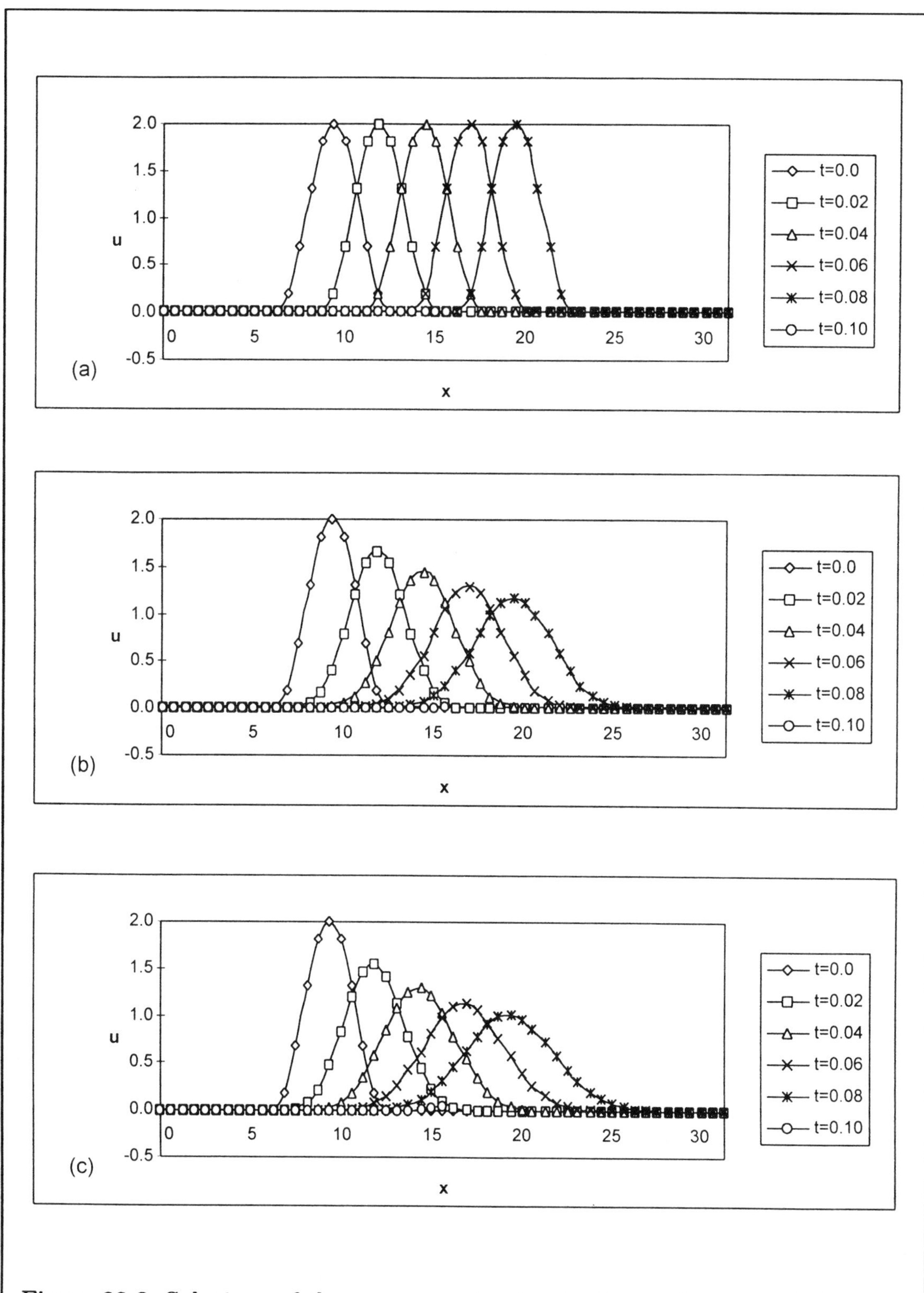

Figure 22-3. Solutions of the wave equation by the first-order upwind scheme.

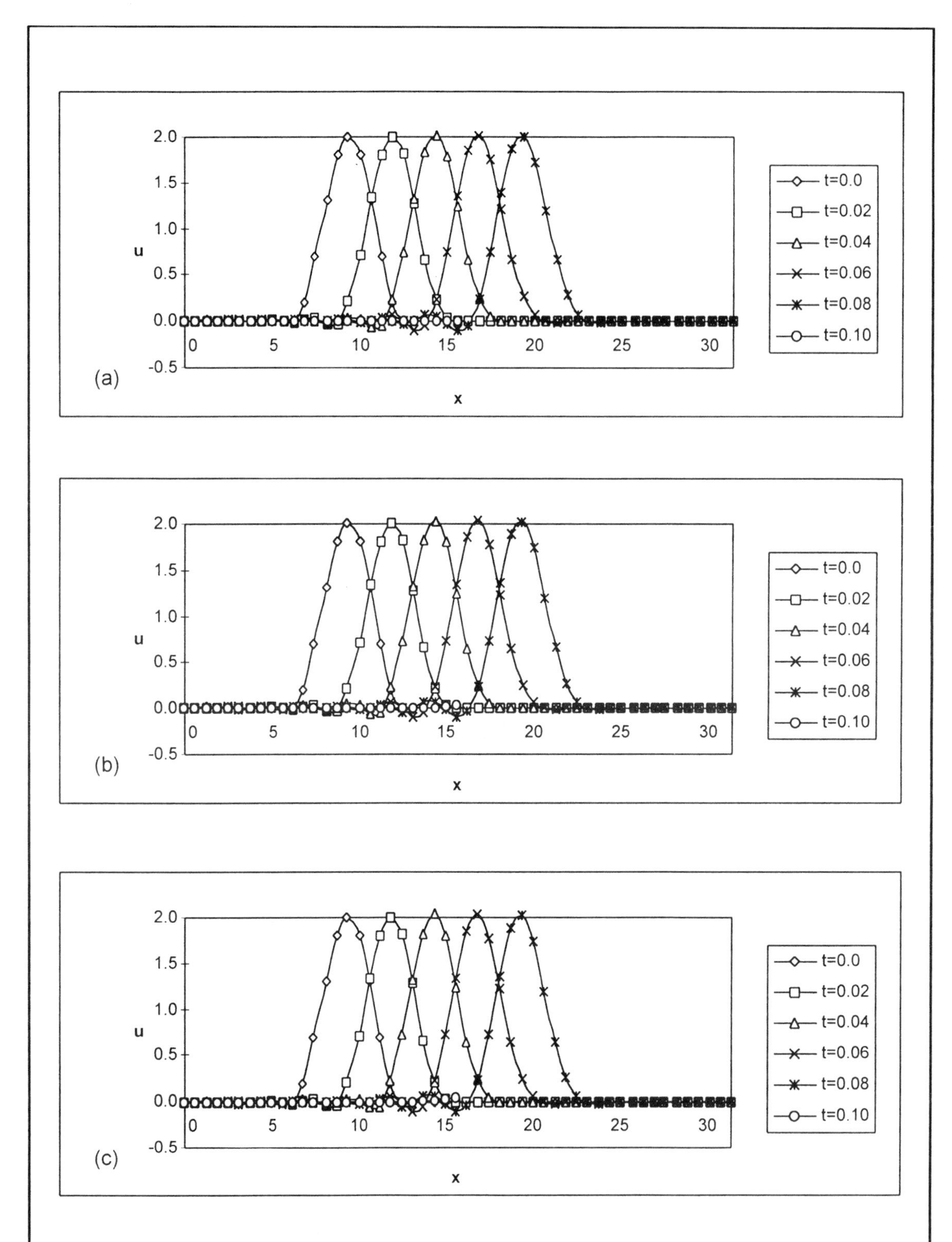

Figure 22-4. Solutions by the fourth-order Runge-Kutta scheme with central finite difference approximation of the convective term.

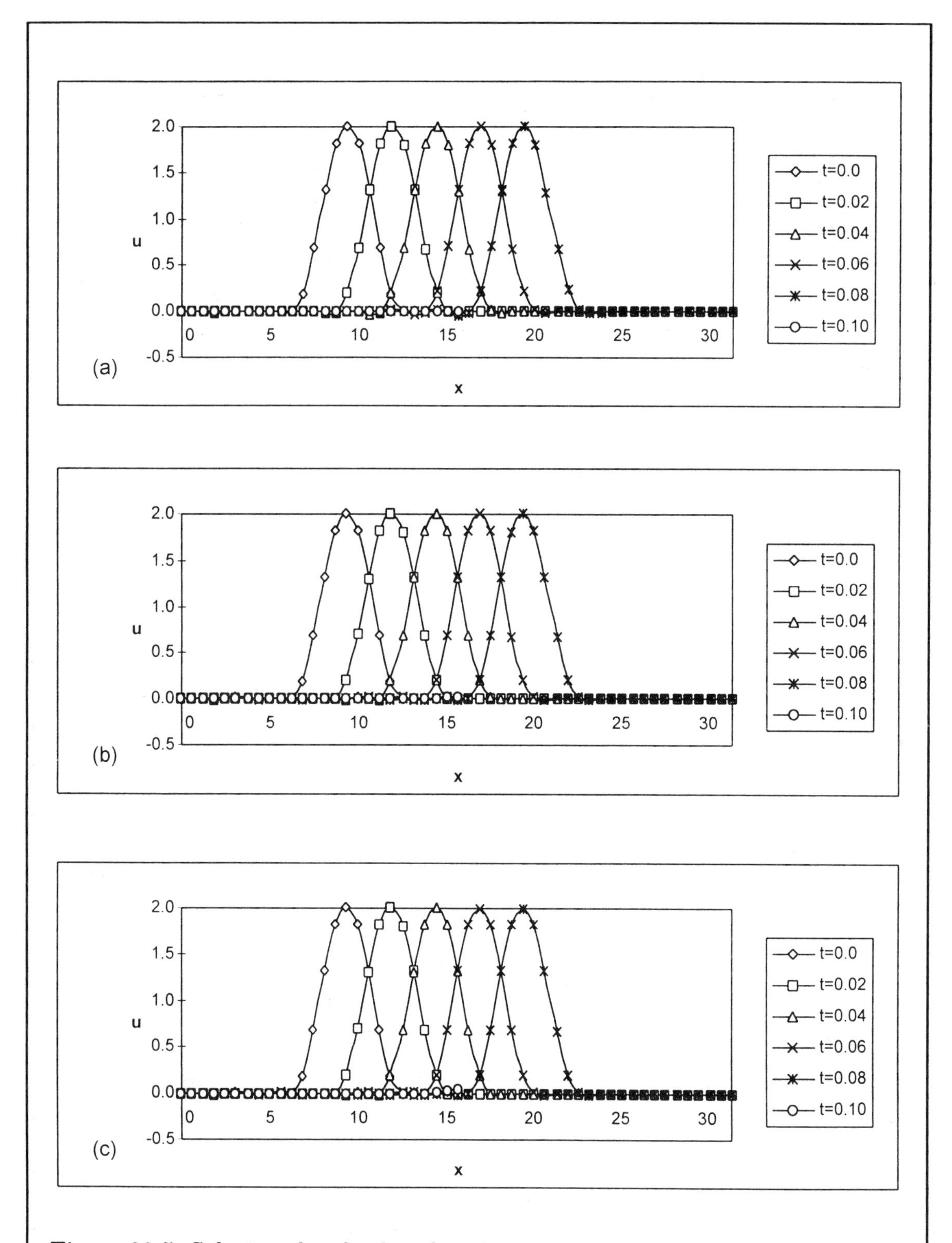

Figure 22-5. Solutions by the fourth-order Runge-Kutta scheme with central compact finite difference approximation of the convective term.

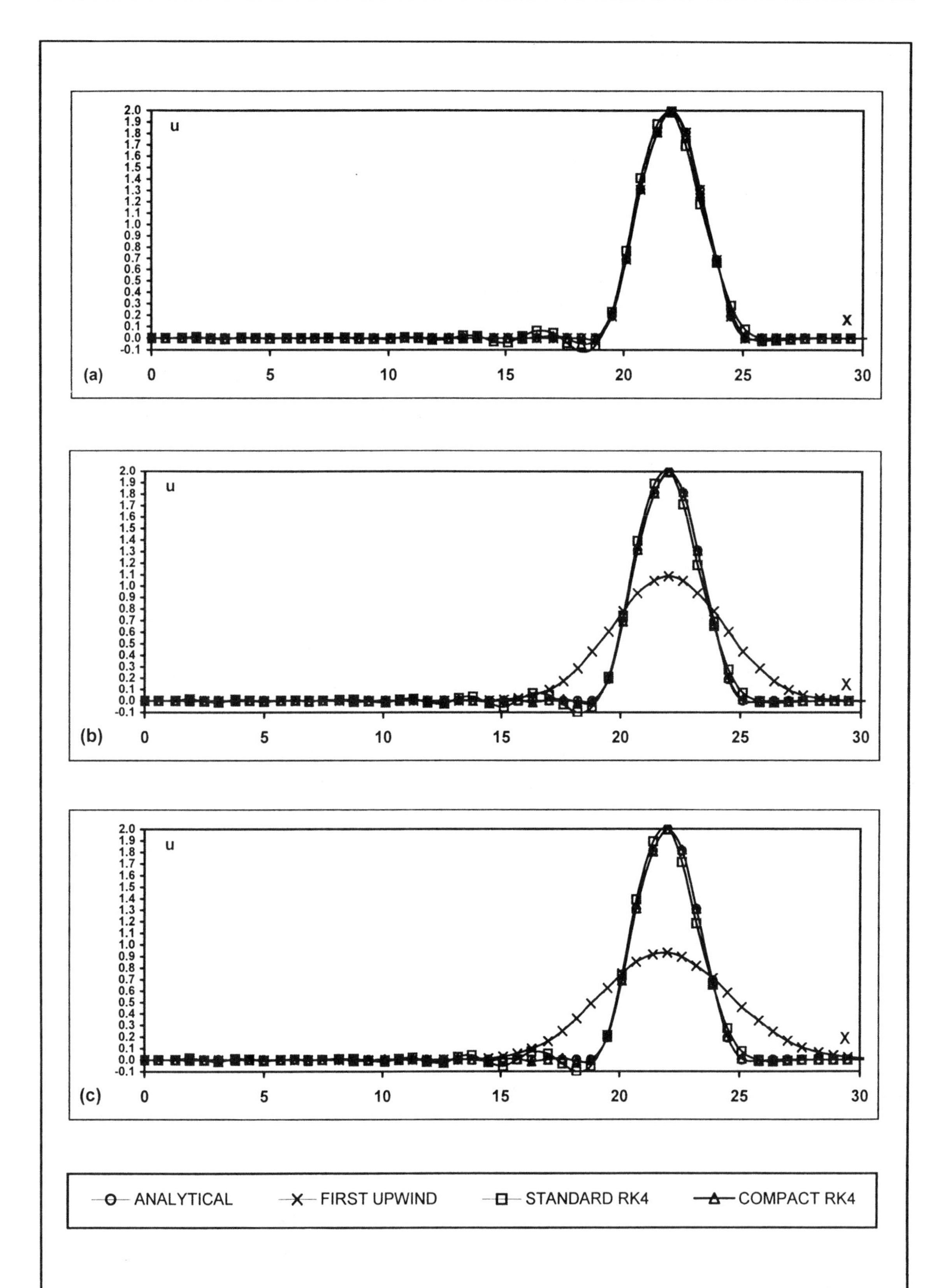

Figure 22-6. Comparison of the solution of the first-order wave equation by several schemes.

22.6 Problems

22.1 Consider a wave which is propagating within a domain with the following initial distribution and shown in Figure P22.1.

$$u(x,0) = 0 \qquad 0 \le x \le 2\pi$$

$$u(x,0) = 0.5\left\{\sin\left[(3x - 7\pi)/2\right] + 1\right\} \qquad 2\pi \le x \le 8\pi/3$$

$$u(x,0) = \sin\left[3(3\pi - x)/2\right] \qquad 8\pi/3 \le x \le 10\pi/3$$

$$u(x,0) = 0.5\left\{\sin\left[(3x - 11\pi)/2\right] - 1\right\} \qquad 10\pi/3 \le x \le 4\pi$$

$$u(x,0) = 0 \qquad 4\pi \le x \le 10\pi$$

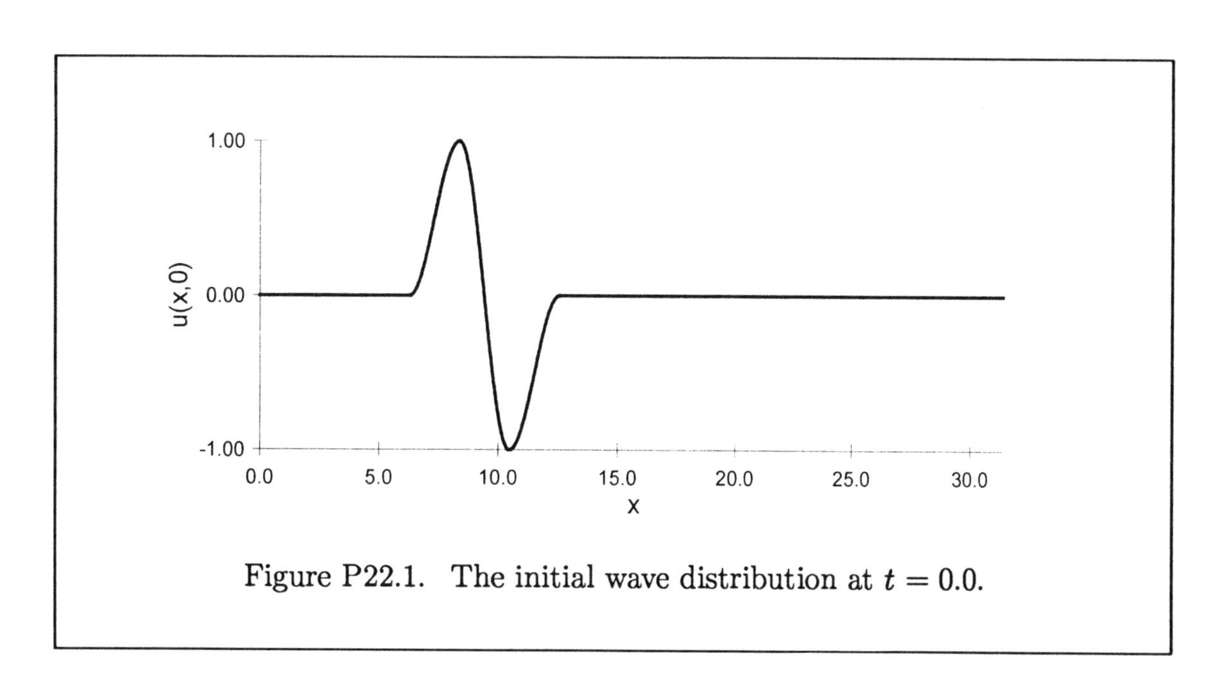

Figure P22.1. The initial wave distribution at $t = 0.0$.

The governing wave equation is given by

$$\frac{\partial u}{\partial t} = -a\frac{\partial u}{\partial x}$$

where a is specified as 40π m/s, and the domain of solution is $0 \le x \le 10\pi$. Select a spatial step of $\Delta x = (10\pi/IM - 1)m$ where $IM = 51$, and temporal steps of

(a) $\Delta t = 0.005$ sec , $c = 1.0$

(b) $\Delta t = 0.0025$ sec , $c = 0.5$

(c) $\Delta t = 0.00125$ sec , $c = 0.25$

Obtain the solution up to $t = 0.1$ sec by the following schemes:

(I) Explicit first upwind scheme

(II) Modified fourth-order Runge-Kutta scheme

(III) Modified fourth-order Runge-Kutta scheme with central compact finite difference approximations

Plot the solutions at intervals of 0.02 sec up to 0.1 sec for all three schemes and for the three Courant numbers.

22.2 Repeat Problem 22.1 for the following initial velocity distribution given by

$$u(x,0) = 2\exp\left[-\left(\frac{x-3\pi}{\pi/2}\right)^2\right]$$

and shown in Figure P22.2.

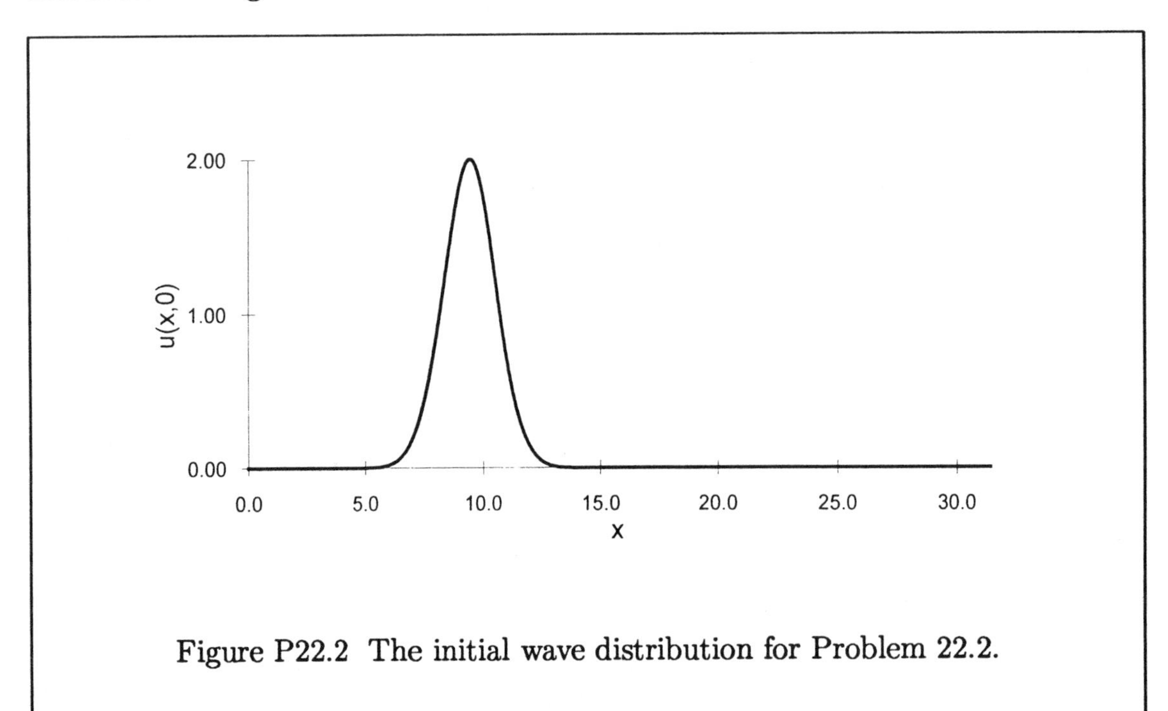

Figure P22.2 The initial wave distribution for Problem 22.2.

Chapter 23

Large Eddy Simulation and Direct Numerical Simulation

23.1 Introductory Remarks

Computation of turbulent flows has always been challenging and remains so at the present and in the foreseeable future. However, since most applications involving fluids include turbulence, a push toward the solution of turbulent flow has been a continuous endeavour. As discussed previously, essentially three avenues are available. First is the solution by the Reynolds Averaged Navier-Stokes (RANS) equation complemented by a turbulence model. Several turbulence models are available for this purpose. These models are continuously evolving (modified and improved) in order to increase their accuracy and range of applications. However, since all scales of turbulence must be modeled in the RANS equations, a model which is capable of predicting turbulence over a wide range of flow conditions and geometries has been illusive. Nonetheless, the RANS approach for computation of turbulent flows is very practical, and the resulting solution is relatively accurate if an appropriate turbulence model is used.

The second approach is the Large Eddy Simulation (LES), and the third approach is the Direct Numerical Simulation (DNS). Application of DNS for flowfield analyses and design purposes at the present is not practical due to limitations on available computers of today and due to the enormous amount of time required. Large Eddy Simulation requires less grid points and computation time, typically five to ten percent of CPU time of DNS computations. However, LES computation still requires substantially more CPU time than similar RANS computation. LES and DNS solutions are, however, very valuable as they can be used to improve turbulence models for RANS and to gain better understanding of physics of turbulence. The objective of this chapter is to introduce LES and DNS.

23.2 Large Eddy Simulation

Large Eddy Simulation provides a compromise between DNS, where all scales of turbulence are computed directly from the Navier-Stokes equations, and RANS equations, where all scales of turbulence must be modeled. In Large Eddy Simulation, small-scale turbulence is filtered out from the Navier-Stokes equations, and a model is used to evaluate small scales. The resulting filtered Navier-Stokes equation is solved for the large-scale motion, which is responsible for most of momentum and energy transport. The large scale of motion is highly dependent on the flow conditions and geometries under consideration. The small scales are computed from the turbulence model known as the *subgridscale model*, which in turn influence the large eddies. Since the small-scale eddies are more or less universal and homogeneous, it is postulated that the subgridscale model would be applicable to a wide range of flow regimes and conditions. Recall that the turbulence models for RANS are very much limited on the range of applications, because they attempt to model a wide range of scales and the random motion of eddies with no organized behavior.

In the following sections, descriptions of filtered Navier-Stokes equations for LES computations and typical subgrid models are presented.

23.2.1 Filtered Navier-Stokes Equations

In order to explore the concept of filtering, consider a simple example of central difference approximation expressed as

$$\left.\frac{\partial f}{\partial x}\right|_i = \frac{f_{i+\frac{1}{2}} - f_{i-\frac{1}{2}}}{\Delta x} = \frac{f(x + \Delta x/2) - f(x - \Delta x/2)}{\Delta x} \tag{23-1}$$

This expression simply approximates the derivative at point i as the averaged values of dependent variable at locations $i + \frac{1}{2}$ and $i - \frac{1}{2}$. Therefore, the expression (23-1) can be written as

$$\frac{1}{\Delta x}\left[f(x + \Delta x/2) - f(x - \Delta x/2)\right] = \frac{d}{dx}\left[\frac{1}{\Delta x}\int_{x-\Delta x/2}^{x+\Delta x/2} f(\xi)d\xi\right] \tag{23-2}$$

which may be interpreted as a filtering operator for which any scale smaller than Δx is filtered out. Now, a filtered quantity is defined as

$$\bar{f}_i(x) = \frac{1}{\Delta}\int_{x-\Delta/2}^{x+\Delta/2} f(\xi)d\xi = \int_{x-\Delta/2}^{x+\Delta/2} f(\xi)G(x,\xi)d\xi \tag{23-3}$$

where, in this case, $G(x,\xi) = 1/\Delta$, G is called a filter function, and Δ is the filter width. Typically, a filter over the entire domain is defined such that

$$\bar{f}(x) = \int_D f(\xi)G(x,\xi)d\xi \tag{23-4}$$

It can be shown that if G is a function of $(x - \xi)$, then differentiation and filtering operation commute [23.1]. Now, (23-4) is written as

$$\bar{f}(x) = \int_D f(\xi)G(x - \xi)d\xi \tag{23-5}$$

Some examples of one-dimensional filter functions are:

1. Top hat filter, $G(x - \xi) = \begin{cases} \dfrac{1}{\Delta} & \text{if } |\, x - \xi \,| < \Delta/2 \\ 0 & \text{otherwise} \end{cases}$

2. Gaussian filter, $G(x - \xi) = \left(\dfrac{6}{\pi\Delta^2}\right)^{3/2} \exp\left[- 6\dfrac{|\, x - \xi \,|^2}{\Delta^2}\right]$

The concept of filtering is extended to three dimensions by the following

$$\bar{f}(x_{i,t}) = \int_D f(\xi_i)G(x_i - \xi_i)d^3\xi_i \tag{23-6}$$

where $x_i = x, y, z$ and $\xi_i = \xi, \eta, \zeta$

Now, define any fluctuation at scales smaller than grid scale Δ as the subgrid scale (SGS) or unresolved quantity, and denote it by a prime. As defined previously by (23-4), the filtered or resolved quantity is denoted by an overbar. Thus,

$$f = \bar{f} + f' \tag{23-7}$$

With the definitions of the resolved and unresolved quantities completed, the Navier-Stokes equations are now modified to yield the Filtered Navier-Stokes (FNS) equations. The FNS equations govern the evolution of large-scale eddies. It will be given for an incompressible flow initially, and, subsequently it will be extended to compressible flows.

After the application of a filtering process, the incompressible Navier-Stokes equations become

$$\frac{\partial \bar{u}_i}{\partial x_i} = 0 \tag{23-8}$$

$$\frac{\partial \bar{u}_i}{\partial t} + \frac{\partial}{\partial x_j}(\bar{u}_i \bar{u}_j) = -\frac{1}{\rho}\frac{\partial \bar{p}}{\partial x_i} + \nu \frac{\partial^2 \bar{u}_i}{\partial x_j \partial x_j} + \frac{\partial \tau_{ij}}{\partial x_j} \tag{23-9}$$

where τ_{ij} is the subgridscale stress term which represents the effect of small scales. The subgrid scale stress is given by

$$\tau_{ij} = \bar{u}_i \bar{u}_j - \overline{u_i u_j} \tag{23-10}$$

which, with the substitution of

$$\bar{u}_i = u_i - u_i' \tag{23-11}$$

can be written as

$$\begin{aligned}\tau_{ij} &= \bar{u}_i\bar{u}_j - \overline{\bar{u}_i\bar{u}_j} - \overline{\bar{u}_i u'_j} - \overline{u'_j\bar{u}_i} - \overline{u'_i u'_j} \\ &= L_{ij} + C_{ij} + R_{ij} \qquad (23\text{-}12)\end{aligned}$$

The terms L_{ij}, C_{ij}, and R_{ij} are defined as follows:

$L_{ij} = \bar{u}_i\bar{u}_j - \overline{\bar{u}_i\bar{u}_j}$ is called *Leonard stress*,

$C_{ij} = -(\overline{\bar{u}_i u'_j} - \overline{u'_j\bar{u}_i})$ is called *cross term stress*, and

$R_{ij} = -\overline{u'_i u'_j}$ is called the *subgridscale Reynolds stress*

Observe that the Leonard stress involves only the resolved quantities, and therefore it can be explicitly computed. Furthermore, note that this term represents the interaction of resolved scales which contribute and affect subgrid scales. The cross term stress and the SGS Reynolds stress involve unresolved quantities and must be modeled. The cross stress term represents the interaction of resolved and unresolved scales, whereas the SGS Reynolds stress represents the interaction of unresolved scales.

It is important to note that a filtered quantity represented by an overbar, that is, $\bar{f}$, is different from the averaging process used in the RANS equation, in particular $\bar{\bar{f}} \neq \bar{f}$.

It is common to rearrange Equation (23-9) and rewrite it as

$$\begin{aligned}\frac{\partial \bar{u}_i}{\partial t} + \frac{\partial}{\partial x_j}(\bar{u}_i\bar{u}_j) &= -\frac{1}{\rho}\frac{\partial}{\partial x_i}\left(\bar{p} - \frac{1}{3}\rho\tau_{kk}\right)\delta_{ij} \\ &\quad + \nu\frac{\partial^2 \bar{u}_i}{\partial x_j \partial x_j} + \frac{\partial}{\partial x_j}\left(\tau_{ij} - \frac{1}{3}\tau_{kk}\delta_{ij}\right) \\ &= -\frac{1}{\rho}\frac{\partial p^+}{\partial x_j} + \nu\frac{\partial^2 \bar{u}_i}{\partial x_j \partial x_j} + \frac{\partial \tau^+_{ij}}{\partial x_j} \qquad (23\text{-}13)\end{aligned}$$

where a modified pressure is defined as

$$p^+ = \bar{p} - \frac{1}{3}\rho\tau_{kk} \qquad (23\text{-}14)$$

and

$$\tau^+_{ij} = \tau_{ij} - \frac{1}{3}\tau_{kk}\delta_{ij} \qquad (23\text{-}15)$$

Extension of the FNS equations to compressible flow is straightforward except that Favre-filtering is used. As in the case of Favre- (or mass-) averaged Navier-Stokes equations, this approach is taken in order to prevent introduction of subgridscale terms into the continuity equation. A Favre-filtered quantity is defined as

$$\tilde{f} = \frac{\overline{\rho f}}{\bar{\rho}} \tag{23-16}$$

Now, the Favre-filtered Navier-Stokes equations are written as

$$\frac{\partial \bar{\rho}}{\partial t} + \frac{\partial}{\partial x_i}(\bar{\rho}\tilde{u}_i) = 0 \tag{23-17}$$

$$\frac{\partial}{\partial t}(\bar{\rho}\tilde{u}_i) + \frac{\partial}{\partial x_j}(\bar{\rho}\tilde{u}_i\tilde{u}_j) + \frac{\partial \tilde{p}}{\partial x_i} = \frac{\partial \tilde{\tau}_{ij}}{\partial x_j} + \frac{\partial \tau_{ij}}{\partial x_j} \tag{23-18}$$

As in the case of incompressible flow, Equation (23-18) is rearranged as

$$\frac{\partial}{\partial t}(\bar{\rho}\tilde{u}_i) + \frac{\partial}{\partial x_j}(\bar{\rho}\tilde{u}_i\tilde{u}_j) + \frac{\partial \tilde{p}}{\partial x_i}\delta_{ij} - \frac{1}{3}\frac{\partial \tau_{kk}}{\partial x_i}\delta_{ij} = \frac{\partial \tilde{\tau}_{ij}}{\partial x_j} + \frac{\partial \tau_{ij}}{\partial x_j} - \frac{1}{3}\frac{\partial \tau_{kk}}{\partial x_j}\delta_{ij} \tag{23-19}$$

or

$$\frac{\partial}{\partial t}(\bar{\rho}\tilde{u}_i) + \frac{\partial}{\partial x_j}(\bar{\rho}\tilde{u}_i\tilde{u}_j) + \frac{\partial}{\partial x_i}\left(\tilde{p} - \frac{1}{3}\tau_{kk}\right)\delta_{ij} = \frac{\partial \tilde{\tau}_{ij}}{\partial x_j} + \frac{\partial}{\partial x_j}\left(\tau_{ij} - \frac{1}{3}\tau_{kk}\delta_{ij}\right) \tag{23-20}$$

Define a modified pressure p^+ as

$$p^+ = \tilde{p} - \frac{1}{3}\tau_{kk} \tag{23-21}$$

and

$$\tau_{ij}^+ = \tau_{ij} - \frac{1}{3}\tau_{kk}\delta_{ij} \tag{23-22}$$

Then, the momentum equation is written as

$$\frac{\partial}{\partial t}(\bar{\rho}\tilde{u}_i) + \frac{\partial}{\partial x_j}(\bar{\rho}\tilde{u}_i\tilde{u}_j) + \frac{\partial p^+}{\partial x_j}\delta_{ij} = \frac{\partial \tilde{\tau}_{ij}}{\partial x_j} + \frac{\partial \tau_{ij}^+}{\partial x_j} \tag{23-23}$$

where

$$\tilde{\tau}_{ij} = -\frac{2}{3}\mu\frac{\partial \tilde{u}_k}{\partial x_k}\delta_{ij} + \mu\left(\frac{\partial \tilde{u}_i}{\partial x_j} + \frac{\partial \tilde{u}_j}{\partial x_i}\right) = \overline{\mu S_{ij}} \tag{23-24}$$

is the filtered stress term and

$$S_{ij} = \frac{\partial u_i}{\partial x_j} + \frac{\partial u_j}{\partial x_i} - \frac{2}{3}\frac{\partial u_k}{\partial x_k}\delta_{ij} \tag{23-25}$$

Furthermore,

$$\tau_{ij} = \bar{\rho}(\tilde{u}_i\tilde{u}_j - \widetilde{u_i u_j}) = \overline{\rho u_i}\,\overline{\rho u_j}/\bar{\rho} - \overline{\rho u_i u_j} \tag{23-26}$$

is the subgrid scale stress.

Note that at this point the subgridscale stress τ_{ij} appearing in Equation (23-20) is an additional unknown which must be modeled. Furthermore, the introduction of τ_{kk} in relations (23-21) and (23-22) and the question of how it may be computed need to be deliberated. These issues will be addressed shortly. For now, the goal is to establish the required set of filtered Navier-Stokes equations for LES of compressible flows. To complete the system of equations, consider the filtered energy equation expressed as

$$\frac{\partial}{\partial t}(\bar{\rho}\tilde{e}_t) + \frac{\partial}{\partial x_i}\left[\overline{(\rho e_t + p)u_i}\right] = \frac{\partial}{\partial x_i}\left(k\frac{\partial T}{\partial x_i}\right) + \frac{\partial}{\partial x_i}(\overline{\mu\, S_{ij} u_j}) \tag{23-27}$$

where

$$\bar{\rho}\tilde{e}_t = \bar{\rho}\tilde{e} + \frac{1}{2}\bar{\rho}(\tilde{u}^2 + \tilde{v}^2 + \tilde{w}^2) - \frac{1}{2}\tau_{kk} \tag{23-28}$$

With the assumption of perfect gas, the total energy given by relation (23-28) can be written as

$$\bar{\rho}\tilde{e}_t = c_v\bar{\rho}\tilde{T} + \frac{1}{2}\bar{\rho}(\tilde{u}^2 + \tilde{v}^2 + \tilde{w}^2) - \frac{1}{2}\tau_{kk} \tag{23-29}$$

which can be rearranged as

$$\bar{\rho}\tilde{e}_t = c_v\bar{\rho}\left(\tilde{T} - \frac{\tau_{kk}}{2c_v\bar{\rho}}\right) + \frac{1}{2}\bar{\rho}(\tilde{u}^2 + \tilde{v}^2 + \tilde{w}^2) \tag{23-30}$$

Consistent with the definition of modified pressure, define a modified temperature as

$$T^+ = \tilde{T} - \frac{\tau_{kk}}{2c_v\bar{\rho}} \tag{23-31}$$

Subsequently

$$\bar{\rho}\tilde{e}_t = c_v\bar{\rho}T^+ + \frac{1}{2}\bar{\rho}(\tilde{u}^2 + \tilde{v}^2 + \tilde{w}^2) \tag{23-32}$$

Consider also the equation of state for a perfect gas which, in terms of the filtered quantities, is written as

$$\bar{p} = \bar{\rho}R\tilde{T} \tag{23-33}$$

In order to write the equation of state given by (23-33) in terms of modified pressure and modified temperature, consider the following

$$\bar{p} - \frac{1}{3}\tau_{kk} = \bar{\rho}R\tilde{T} - \frac{1}{3}\tau_{kk} - \frac{\tau_{kk}}{2c_v\bar{\rho}}\bar{\rho}R + \frac{\tau_{kk}}{2c_v\bar{\rho}}\bar{\rho}R =$$

$$\bar{\rho}R\left(\tilde{T} - \frac{\tau_{kk}}{2c_v\bar{\rho}}\right) + \tau_{kk}\left(\frac{R}{2c_v} - \frac{1}{3}\right)$$

or

$$p^+ = \bar{\rho} R T^+ + \left(\frac{R}{2c_v} - \frac{1}{3} \right) \tau_{kk} \tag{23-34}$$

Now the energy equation given by (23-27) is written in terms of the modified pressure. Consider the second term given by $\overline{(\rho\, e_t + p) u_i}$ and rewrite it as

$$\begin{aligned} \overline{(\rho\, e_t + p) u_i} &= \overline{(\rho\, e_t + p) u_i} - (\bar{\rho}\tilde{e}_t + p^+)\tilde{u}_i + (\bar{\rho}\tilde{e}_t + p^+)\tilde{u}_i \\ &= -\, Q_i + (\bar{\rho}\tilde{e}_t + p^+)\tilde{u}_i \end{aligned} \tag{23-35}$$

where

$$Q_i = -\overline{(\rho\, e_t + p) u_i} + (\bar{\rho}\tilde{e}_t + p^+)\tilde{u}_i \tag{23-36}$$

is the subgrid heat flux. Thus, the energy equation is now written as

$$\frac{\partial}{\partial t}(\bar{\rho}\tilde{e}_t) + \frac{\partial}{\partial x_i}\left[(\bar{\rho}\tilde{e}_t + p^+)\tilde{u}_i \right] = \frac{\partial Q_i}{\partial x_i} + \frac{\partial}{\partial x_i}\left(\overline{k \frac{\partial T}{\partial x_i}} \right) + \frac{\partial}{\partial x_i}\left(\overline{\mu\, u_j\, S_{ij}} \right) \tag{23-37}$$

For a relatively high Reynolds number, typically the following is introduced in the momentum and energy equations, respectively.

$$\overline{\mu\, S_{ij}} = \mu\, \tilde{S}_{ij} \tag{23-38}$$

and

$$\overline{\mu\, u_j\, S_{ij}} = \mu\, \tilde{u}_j\, \tilde{S}_{ij} \tag{23-39}$$

Thus, the system of equations composed of filtered continuity, momentum, and energy equations is written as

$$\frac{\partial \bar{\rho}}{\partial t} + \frac{\partial}{\partial x_i}(\bar{\rho}\tilde{u}_i) = 0 \tag{23-40}$$

$$\frac{\partial}{\partial t}(\bar{\rho}\tilde{u}_i) + \frac{\partial}{\partial x_j}(\bar{\rho}\tilde{u}_i\tilde{u}_j + p^+\delta_{ij}) = \frac{\partial \tau_{ij}^+}{\partial x_j} + \frac{\partial}{\partial x_j}(\mu\, \tilde{S}_{ij}) \tag{23-41}$$

$$\frac{\partial}{\partial t}(\bar{\rho}\tilde{e}_t) + \frac{\partial}{\partial x_i}\left[(\bar{\rho}\tilde{e}_t + p^+)\tilde{u}_i \right] = \frac{\partial Q_i}{\partial x_i} + \frac{\partial}{\partial x_i}\left(\overline{k \frac{\partial T}{\partial x_i}} \right) + \frac{\partial}{\partial x_i}(\mu\, \tilde{u}_j\, \tilde{S}_{ij}) \tag{23-42}$$

These equations are now expanded in Cartesian coordinate and are written in a flux vector formulation by defining

$$Q = \begin{bmatrix} \bar{\rho} \\ \bar{\rho}\tilde{u} \\ \bar{\rho}\tilde{v} \\ \bar{\rho}\tilde{w} \\ \bar{\rho}\tilde{e}_t \end{bmatrix} \tag{23-43}$$

$$E = \begin{bmatrix} \bar{\rho}\tilde{u} \\ \bar{\rho}\tilde{u}^2 + p^+ \\ \bar{\rho}\tilde{u}\tilde{v} \\ \bar{\rho}\tilde{u}\tilde{w} \\ (\bar{\rho}\tilde{e}_t + p^+)\tilde{u} \end{bmatrix} \tag{23-44}$$

$$F = \begin{bmatrix} \bar{\rho}\tilde{v} \\ \bar{\rho}\tilde{v}\tilde{u} \\ \bar{\rho}\tilde{v}^2 + p^+ \\ \bar{\rho}\tilde{v}\tilde{w} \\ (\bar{\rho}\tilde{e}_t + p^+)\tilde{v} \end{bmatrix} \quad (23\text{-}45) \qquad G = \begin{bmatrix} \bar{\rho}\tilde{w} \\ \bar{\rho}\tilde{w}\tilde{u} \\ \bar{\rho}\tilde{w}\tilde{v} \\ \bar{\rho}\tilde{w}^2 + p^+ \\ (\bar{\rho}\tilde{e}_t + p^+)\tilde{w} \end{bmatrix} \quad (23\text{-}46)$$

$$E_v = \begin{bmatrix} 0 \\ \tau_{xx}^+ + \mu\,\tilde{S}_{xx} \\ \tau_{xy}^+ + \mu\,\tilde{S}_{xy} \\ \tau_{xz}^+ + \mu\,\tilde{S}_{xz} \\ Q_x + k\frac{\partial T^+}{\partial x} + \mu\,(\tilde{u}\tilde{S}_{xx} + \tilde{v}\,\tilde{S}_{xy} + \tilde{w}\,\tilde{S}_{xz}) \end{bmatrix} \quad (23\text{-}47)$$

$$F_v = \begin{bmatrix} 0 \\ \tau_{yx}^+ + \mu\,\tilde{S}_{yx} \\ \tau_{yy}^+ + \mu\tilde{S}_{yy} \\ \tau_{yz}^+ + \mu\tilde{S}_{yz} \\ Q_y + k\frac{\partial T^+}{\partial y} + \mu\,(\tilde{u}\tilde{S}_{yx} + \tilde{v}\,\tilde{S}_{yy} + \tilde{w}\,\tilde{S}_{yz}) \end{bmatrix} \quad (23\text{-}48)$$

$$G_v = \begin{bmatrix} 0 \\ \tau_{zx}^+ + \mu\tilde{S}_{zx} \\ \tau_{zy}^+ + \mu\tilde{S}_{zy} \\ \tau_{zz}^+ + \mu\tilde{S}_{zz} \\ Q_z + k\frac{\partial T^+}{\partial z} + \mu\,(\tilde{u}\tilde{S}_{zx} + \tilde{v}\,\tilde{S}_{zy} + \tilde{w}\,\tilde{S}_{zz}) \end{bmatrix} \quad (23\text{-}49)$$

Now the flux vector formulation, similar to that of the Navier-Stokes equation given by (14-1), is written as

$$\frac{\partial Q}{\partial t} + \frac{\partial E}{\partial x} + \frac{\partial F}{\partial y} + \frac{\partial G}{\partial z} = \frac{\partial E_v}{\partial x} + \frac{\partial F_v}{\partial y} + \frac{\partial G_v}{\partial z} \quad (23\text{-}50)$$

The subgridscale terms τ_{ij}^+ and Q_i are typically expressed in terms of the eddy viscosity and eddy diffusivity similar to that in RANS equations and is presented next.

23.2.2 Subgridscale Models

At this point, certain parallelisms between LES and RANS can be realized. Since a relatively strong background on RANS and turbulence models has already been established, continuous reference to RANS and turbulence models will be made as options for LES and subgridscale models are explored.

As in the case of RANS where turbulence quantities are written in terms of eddy viscosity and eddy diffusivity, the subgridscale terms are also expressed in terms of eddy viscosity and eddy diffusivity. For most applications in either RANS or LES, a model for eddy viscosity is developed, and, subsequently, eddy diffusivity is determined by the introduction of a turbulent Prandtl number.

Subgridscale models similar to turbulence models vary in sophistication from simple algebraic models to one-equation models to multi-equation models.

Several concepts with regard to small scales of turbulence were identified in Chapter One which are relevant to the development of SGS models and are reviewed at this point. The small scales are more uniform and isotropic than large scales. Therefore, modeling of small scales by simple algebraic expression tends to represent the physics fairly well and is applicable to a wider range of flow conditions. Furthermore, subgridscale stress contributes a fraction of total stress, and, therefore, small errors in modeling do not substantially effect the overall accuracy.

In addition to assumptions of isotropic and uniformity of small scales, the assumption of equilibrium is also imposed for most applications. The assumption of equilibrium is referred to as a condition where the energy contained in large eddies is transferred to smaller eddies, and, ultimately, all are dissipated into heat at the level of smallest eddies (subgridscale) due to molecular viscosity. Although, on the average, the transfer of energy takes place from resolved scales to unresolved scales, the reverse can occur (known as *backscatter*), which is not accounted for when equilibrium assumption is used.

Based on these discussions, algebraic models appear to be good candidates for subgridscale terms. Besides their simplicity, they require a minimum amount of computational effort. In the following sections, algebraic SGS models are introduced.

23.2.2.1 Eddy Viscosity

The subgridscale stress is related to the large-scale strain rate tensor by introduction of eddy viscosity and is written in a similar form as the laminar stress. For an incompressible flow, the subgridscale stress term is written as

$$\tau_{ij}^{+} = \tau_{ij} - \frac{1}{3}\tau_{kk}\,\delta_{ij} = 2\nu_t\,\bar{S}_{ij} \tag{23-51}$$

where

$$\bar{S}_{ij} = \frac{1}{2}\left(\frac{\partial \bar{u}_i}{\partial x_j} + \frac{\partial \bar{u}_j}{\partial x_i}\right) \tag{23-52}$$

The subgridscale stress term for compressible flow is written as

$$\tau_{ij}^{+} = \tau_{ij} - \frac{1}{3}\tau_{kk}\,\delta_{ij} = \bar{\rho}\,\nu_t\,\tilde{S}_{ij} \tag{23-53}$$

and $\tilde{S}_{ij}$ is given by

$$\tilde{S}_{ij} = \frac{\partial \tilde{u}_i}{\partial x_j} + \frac{\partial \tilde{u}_j}{\partial x_i} - \frac{2}{3}\frac{\partial \tilde{u}_k}{\partial x_k}\,\delta_{ij} \tag{23-54}$$

The subgrid heat flux is written in terms of eddy diffusivity and, subsequently, in terms of turbulent Prandtl number as follows.

$$Q_i = \frac{k_t}{c_p}\frac{\partial T^+}{\partial x_i} = \bar{\rho}\,\frac{\nu_t}{Pr_t}\frac{\partial T^+}{\partial x_i} \tag{23-55}$$

Observe that, just as in the RANS equations, the eddy conductivity k_t is replaced by terms of eddy viscosity and turbulent Prandtl number. Again, just as in the RANS equations, several models are available to determine ν_t.

23.2.2.1.1 Smagorinsky Model

The first subgridscale stress model was developed by Smagorinsky [23.2] in the 1960's. It is a simple model which provides reasonable results for some applications. This model is still commonly used in LES due to its simplicity. Several modifications have been proposed to improve its prediction capabilities. One such attempt is the dynamic model which will be discussed in the next section.

The Smagorinsky model is developed based on the assumption of equilibrium. The eddy viscosity is written to be proportional to a length scale ℓ.

The subgridscale stress is written in terms of eddy viscosity as

$$\tau_{ij} - \frac{1}{3}\,\tau_{kk}\delta_{ij} = 2\nu_t\,\bar{S}_{ij} \tag{23-56}$$

where, for an incompressible flow, the large-scale strain rate is given by

$$\bar{S}_{ij} = \frac{1}{2}\left(\frac{\partial \bar{u}_i}{\partial x_j} + \frac{\partial \bar{u}_j}{\partial x_i}\right) \tag{23-57}$$

and in terms of the velocity scale q_{SGS} as follows

$$\nu_t \sim \ell\, q_{SGS} \tag{23-58}$$

The length scale is taken as the filter width Δ, and the velocity scale $q_{SGS} \sim \ell|\bar{S}|$, where

$$|\bar{S}| = \sqrt{2\bar{S}_{ij}\,\bar{S}_{ij}} \tag{23-59}$$

Finally,

$$\nu_t \sim \Delta^2\,|\bar{S}| \tag{23-60}$$

which is written as an equation by introduction of the Smagorinsky constant as

$$\nu_t = c_s^2 \Delta^2 |\bar{S}| \tag{23-61}$$

The Smagorinsky constant can be approximated in terms of Kolmogorov constant c_k by the following

$$c_s \approx \frac{1}{\pi} \left(\frac{3}{2} c_k \right)^{-3/4}$$

where, if $c_k = 1.4$, then $c_s = 0.18$. This value of Smagorinsky constant is not universal. In fact, similar to the constants used in the algebraic turbulence models, it's value would vary. However, for most applications, a constant in the range of $0.1 < c_s < 0.24$ have been used.

The Smagorinsky model and its application are simple and provide reasonable results for some flows. However, there are several drawbacks to the model.

- First, as mentioned above, the value of Smagorinsky constant is not universal, and its value is different for different types of flows.

- Second, due to the imposed equilibrium assumption, the subgridscale stress always decreases the energy of flow, and, therefore, backscatter which may be present in the flow is omitted.

- Third, it may be necessary to reduce the model constant in the near wall region.

- Fourth, since, for most applications, grid clustering will be used, specification of filter width Δ may be difficult. Among several options available, one may consider $\Delta = (\Delta x^2 + \Delta y^2 + \Delta z^2)^{1/2}$ or $\Delta = (\Delta x \Delta y \Delta z)^{1/3}$.

An improvement to the model can be made where the value of constant c_s would vary locally, thereby reducing or increasing eddy viscosity in locations as required. This concept is the basis of the dynamic model which is provided next.

23.2.2.1.2 Dynamic Model

Several potential problems with the original Smagorinsky model may develop in LES of near wall flows, shear flows, or in transition regions as discussed in the previous section. These difficulties are due to the selection of a constant value for Smagorinsky coefficient. From physical observations, it is known that, for example, backscatter occurs in some flows which in fact can be significant. Furthermore, the subgridscale stress must possess assymptotic behavior near the wall and vanish at the wall. To account for these physical considerations and to overcome the difficulties associated with a constant value of c_s, a dynamic model is proposed by Germano [23.3]. In this model, the coefficient c_s is allowed to vary locally to adjust the level of eddy viscosity in the flowfield. The procedure is initiated by specifying a value of c_s, and subsequently after each time level or in practice after several time levels, the flowfield is examined, and a new value of c_s at each point is determined. Therefore, the value of the coefficient c_s is continuously modified as solution proceeds in time. The scheme is similar in principle to adaptive grid, where the grid system evolves as the solution proceeds in time.

The dynamic model is developed by the introduction of a test filter with a length scale larger than that of the length scale of the original filter, typically by a factor of two. Stresses in this range are referred to as subtest-scale (STS) stresses and are modeled similar to the SGS stresses. The test filter is defined by $\hat{G}$, and a tophat "^" is used to identify the associated test-filtered quantities. After the test filter is applied to the FNS equations, the STS stress is developed and is given by

$$\hat{\bar{\tau}}_{ij} = \hat{\bar{u}}_i \hat{\bar{u}}_j - \widehat{\overline{u_i u_j}} \tag{23-62}$$

and the resolved stress is

$$\hat{L}_{ij} = \hat{\bar{u}}_i \hat{\bar{u}}_j - \widehat{\bar{u}_i \bar{u}_j} \tag{23-63}$$

Recall that the SGS stress as given by (23-10) is

$$\tau_{ij} = \bar{u}_i \bar{u}_j - \overline{u_i u_j} \tag{23-64}$$

When the test filter is applied to (23-64), one obtains

$$\hat{\tau}_{ij} = \widehat{\bar{u}_i \bar{u}_j} - \widehat{\overline{u_i u_j}} \tag{23-65}$$

It is then obvious that one can write

$$\hat{L}_{ij} - \hat{\bar{\tau}}_{ij} = -\hat{\tau}_{ij} \tag{23-66}$$

The resolved stress $\hat{L}_{ij}$ can be computed explicitly from the resolved velocity field, whereas the SGS stress and STS stress require modeling.

In the following, the dynamic model is used in conjunction with the Smagorinsky model. In fact, the procedure can be applied to any eddy viscosity model. Now, the Smagorinsky model applied to SGS and STS stresses is written as

$$\hat{\tau}_{ij} - \frac{1}{3} \hat{\tau}_{kk} \delta_{ij} = 2 \widehat{c\,\alpha_{ij}} \tag{23-67}$$

and

$$\hat{\bar{\tau}}_{ij} - \frac{1}{3} \hat{\bar{\tau}}_{kk} \delta_{ij} = 2c\,\beta_{ij} \tag{23-68}$$

where

$$\alpha_{ij} = \bar{\Delta}^2 |\bar{S}| \bar{S}_{ij} \qquad (23\text{-}69)$$

$$\beta_{ij} = \hat{\Delta}^2 |\hat{\bar{S}}| \hat{\bar{S}}_{ij} \qquad (23\text{-}70)$$

Subtracting (23-67) from (23-68), one obtains

$$\hat{\bar{\tau}}_{ij} - \hat{\tau}_{ij} - \frac{1}{3}\left(\hat{\bar{\tau}}_{kk} - \hat{\tau}_{kk}\right)\delta_{ij} = 2c\,\beta_{ij} - \widehat{2c\,\alpha_{ij}} \qquad (23\text{-}71)$$

or

$$\hat{L}_{ij} - \frac{1}{3}\hat{L}_{kk}\,\delta_{ij} = 2c\,\beta_{ij} - \widehat{2c\,\alpha_{ij}} \qquad (23\text{-}72)$$

Typically, c is removed from the filtering operation such that (23-72) is written as

$$\hat{L}_{ij} - \frac{1}{3}\hat{L}_{kk}\,\delta_{ij} = 2c\,(\beta_{ij} - \hat{\alpha}_{ij}) = 2c\,M_{ij} \qquad (23\text{-}73)$$

where

$$M_{ij} = \beta_{ij} - \hat{\alpha}_{ij} \qquad (23\text{-}74)$$

Equation (23-73) yields five independent equations with only one unknown c, that is, the system is overdetermined. One approach to overcome this problem is proposed by Germano, in which Equation (23-73) is contracted by $\bar{S}_{ij}$, resulting in

$$\hat{L}_{ij}\,\bar{S}_{ij} = 2c\,M_{ij}\,\bar{S}_{ij} \qquad (23\text{-}75)$$

from which

$$c = \frac{1}{2}\frac{\hat{L}_{ij}\,\bar{S}_{ij}}{M_{ij}\,\bar{S}_{ij}} \qquad (23\text{-}76)$$

Another approach is proposed by Lilly [23.4], in which the sum of the squares of the residual is minimized. The result is

$$c = \frac{1}{2}\frac{\hat{L}_{ij}\,M_{ij}}{M_{ij}^2} \qquad (23\text{-}77)$$

This formulation is particularly attractive because the formulation given by (23-76) may cause a difficulty in that the denominator could become locally zero or very small such that numerical instability within the solution develops. This difficulty in formulation (23-76) can be removed by the assumption that c is a function of y and t only. This assumption allows the introduction of an averaging scheme taken over the x and z directions, that is, parallel to the wall. This averaging scheme (indicated by $< >$) also addresses another difficulty which may be encountered when either (23-76) or (23-77) is used. It has been observed that the value of c could vary significantly over the domain and may possess negative values over large regions. This may also become a source of instability in the solution. Though a certain value of negative c is desireable in the solution, large values could create serious problems. The negative values of c are interpreted as allowing backscatter in the solution. Introduction of the averaging procedure discussed above tends to eliminate the problem of excessive negative values of c. However, note that, with the introduction of the averaging process over the domain, the model can no longer be considered strictly local.

Now the formulations of c are written in terms of the averaging scheme as follows.

$$c = \frac{1}{2} \frac{< \hat{L}_{ij} \bar{S}_{ij} >}{< M_{ij} \bar{S}_{ij} >} \tag{23-78}$$

and

$$c = \frac{1}{2} \frac{< \hat{L}_{ij} M_{ij} >}{< M_{ij}^2 >} \tag{23-79}$$

In closing this section, it is concluded that the dynamic model overcomes some of the problems associated with eddy viscosity models with constant coefficient. In the process, however, the dynamic model also creates some difficulties as identified above. However, some simple procedures were introduced by which these difficulties can be removed.

23.3 Direct Numerical Simulation

The third category of solution scheme for laminar, transitional, and turbulent flows is Direct Numerical Simulation. This scheme essentially solves the time dependent Navier-Stokes equations and attempts to accommodate all scales of turbulence. There are two major grand challenges that need to be resolved in order for DNS to be a routine computational scheme for design and analysis purposes. The first major obstacle is the limitation of the available resources. Present day computers do not have the memory, operation speed, and data transfer capabilities to conduct a DNS of complex configuration at high Reynolds number.

It was shown in Chapter 20 that the number of grid points required for DNS is proportional to $Re^{9/4}$. For example, a DNS of a complete aircraft will require about one extaflop, that is 10^{18} flops. The projected computer performance is approximately 10^{15} (operations per second) to be achieved in about 2010. One approach to gain higher performance and increase memory is to perform DNS on multi-processor, parallel computers. In this approach, the domain of solution is divided among several processors, where information is transferred between the processors. And, of course, consideration of data transfer and the transfer speed is crucial. However, multi-processor parallel computations have shown great potential in performance of DNS for turbulent flows over large domain at high Reynolds numbers.

The second challenge is the development of a solution algorithm which is free of significant numerical error. This aspect of the problem includes the grid system, numerical scheme, and boundary conditions.

The grid system must be developed by a higher-order grid generation scheme such as to provide a high quality grid, that is, a grid for which there is continuity in the curvature and the metrics obtained by high-order approximations, for example, fourth-order or higher. An example of such grid generation is the fourth-order elliptic grid generation scheme.

The development of numerical schemes must include higher-order approximation of the derivatives. The fourth-order Runge-Kutta scheme can be used for time integration, whereas typically sixth-order compact finite difference approximation is used for the convection and diffusion terms. The third element of the solution algorithm which is crucial is the treatment of boundary conditions. The sponge layer type boundary treatment is usually implemented for the inflow/outflow boundaries, and the standard no-slip boundary condition is applied at the solid surface.

Some examples of DNS of transitional/turbulent flows which have been reported in the literature are provided below. As the technology matures, the envelope and the range of applications will increase to include higher Reynolds numbers and more complex geometries. The following list is a very limited example of DNS and by no means is complete.

- Compressible free shear flow. Several computations are performed at convective Mach number (M_c) of 0.38 ($M_1 = 1.5$, $M_2 = 1.5$), M_c of 0.4 ($M_1 = 2.0$, $M_2 = 1.2$), and M_c of 0.8 ($M_1 = 4.0$, $M_2 = 2.4$). The Reynolds number range was between 100 and 500, where $Re = \rho_1(u_1 - u_2)\delta_\omega/\mu_1$. The subscripts 1 and 2 refer to high and low speed streams, and δ_ω is the initial vorticity thickness [23.5].

- Transitional flow at Mach 0.5 and Re_L of 160,000 over a Joukowsky airfoil. The effect of suction/blowing is considered as a means for controlled transition [23.6].

- Flow transition at Mach 1.6 over a Joukowsky airfoil. The Reynolds number based on the half-thickness of the airfoil h was $Re_h = 15000$ [23.7].

- Compressible flat plate boundary layer flow at Mach 2.25. The Reynolds number based on the inlet conditions was 635000/in [23.8].

- Fully turbulent channel flow. Effect of wall injection and suction was investigated. The Reynolds number based on the channel half width and averaged friction velocity was 150 [23.9].

- Supersonic turbulent boundary-layer flow over a flat plate and a compression ramp. The compressible, turbulent flowfield over a flat plate at Mach numbers of 3, 4.5, and 6, and at momentum thickness Reynolds numbers of 3015, 2618, and 2652 were investigated. In addition, flowfield over an 18 degree compression ramp at Mach 3 and Reynolds number of 5000 was considered [23.11].

APPENDIX J:

Transformation of Turbulence Models from Physical Space to Computational Space

The details of the coordinate transformation for turbulence models are provided in this appendix. Since the terms for most models in each category, e.g., one-equation models, are similar, mathematical steps will be carried out for a typical model only. Extension to similar turbulence models is straightforward.

J.1 Baldwin-Barth Turbulence Model

To better manage the mathematical work, each term in the equation will be considered separately. Again, using the notation R for $\nu \overline{Re_T}$, the left-hand side of (21-86) is dR/dt, which is expanded in Cartesian coordinates to yield

$$\frac{dR}{dt} = \frac{\partial R}{\partial t} + \vec{V} \cdot \nabla R = \frac{\partial R}{\partial t} + u\frac{\partial R}{\partial x} + v\frac{\partial R}{\partial y}$$

Now, using the Cartesian derivatives (21-88) and (21-89), the convective term becomes

$$\begin{aligned} u\frac{\partial R}{\partial x} + v\frac{\partial R}{\partial y} &= u\left(\xi_x \frac{\partial R}{\partial \xi} + \eta_x \frac{\partial R}{\partial \eta}\right) + v\left(\xi_y \frac{\partial R}{\partial \xi} + \eta_y \frac{\partial R}{\partial \eta}\right) \\ &= (\xi_x u + \xi_y v)\frac{\partial R}{\partial \xi} + (\eta_x u + \eta_y v)\frac{\partial R}{\partial \eta} = U\frac{\partial R}{\partial \xi} + V\frac{\partial R}{\partial \eta} \end{aligned}$$

The diffusion term in Equation (21-86) involves the Laplacian $\nabla^2 R$ and $\nabla \nu_t \cdot \nabla R$. Each term is treated separately, and details are as follow.

Using the expressions given by (21-90) and (21-91), one has

$$\nabla^2 R = \frac{\partial^2 R}{\partial x^2} + \frac{\partial^2 R}{\partial y^2} = \xi_x^2 \frac{\partial^2 R}{\partial \xi^2} + 2\xi_x \eta_x \frac{\partial^2 R}{\partial \xi \partial \eta} + \eta_x^2 \frac{\partial^2 R}{\partial \eta^2} +$$

$$+\xi_x\left(\frac{\partial\xi_x}{\partial\xi}\frac{\partial R}{\partial\xi}+\frac{\partial\eta_x}{\partial\xi}\frac{\partial R}{\partial\eta}\right)+\eta_x\left(\frac{\partial\xi_x}{\partial\eta}\frac{\partial R}{\partial\xi}+\frac{\partial\eta_x}{\partial\eta}\frac{\partial R}{\partial\eta}\right)$$

$$+\xi_y^2\frac{\partial^2 R}{\partial\xi^2}+2\xi_y\eta_y\frac{\partial^2 R}{\partial\xi\partial\eta}+\eta_y^2\frac{\partial^2 R}{\partial\eta^2}$$

$$+\xi_y\left(\frac{\partial\xi_y}{\partial\xi}\frac{\partial R}{\partial\xi}+\frac{\partial\eta_y}{\partial\xi}\frac{\partial R}{\partial\eta}\right)+\eta_y\left(\frac{\partial\xi_y}{\partial\eta}\frac{\partial R}{\partial\xi}+\frac{\partial\eta_y}{\partial\eta}\frac{\partial R}{\partial\eta}\right)$$

Rearrange terms,

$$\begin{aligned}
\nabla^2 R &= \left(\xi_x^2+\xi_y^2\right)\frac{\partial^2 R}{\partial\xi^2}+2\left(\xi_x\eta_x+\xi_y\eta_y\right)\frac{\partial^2 R}{\partial\xi\partial\eta}+\left(\eta_x^2+\eta_y^2\right)\frac{\partial^2 R}{\partial\eta^2}+\\
&\quad+\xi_x\left(\frac{\partial\xi_x}{\partial\xi}\frac{\partial R}{\partial\xi}+\frac{\partial\eta_x}{\partial\xi}\frac{\partial R}{\partial\eta}\right)+\eta_x\left(\frac{\partial\xi_x}{\partial\eta}\frac{\partial R}{\partial\xi}+\frac{\partial\eta_x}{\partial\eta}\frac{\partial R}{\partial\eta}\right)\\
&\quad+\xi_y\left(\frac{\partial\xi_y}{\partial\xi}\frac{\partial R}{\partial\xi}+\frac{\partial\eta_y}{\partial\xi}\frac{\partial R}{\partial\eta}\right)+\eta_y\left(\frac{\partial\xi_y}{\partial\eta}\frac{\partial R}{\partial\xi}+\frac{\partial\eta_y}{\partial\eta}\frac{\partial R}{\partial\eta}\right)\\
&= \left(\xi_x^2+\xi_y^2\right)\frac{\partial^2 R}{\partial\xi^2}+2(\xi_x\eta_x+\xi_y\eta_y)\frac{\partial^2 R}{\partial\xi\partial\eta}+(\eta_x^2+\eta_y^2)\frac{\partial^2 R}{\partial\eta^2}\\
&\quad+\left(\xi_x\frac{\partial\xi_x}{\partial\xi}+\eta_x\frac{\partial\xi_x}{\partial\eta}+\xi_y\frac{\partial\xi_y}{\partial\xi}+\eta_y\frac{\partial\xi_y}{\partial\eta}\right)\frac{\partial R}{\partial\xi}\\
&\quad+\left(\xi_x\frac{\partial\eta_x}{\partial\xi}+\eta_x\frac{\partial\eta_x}{\partial\eta}+\xi_y\frac{\partial\eta_y}{\partial\xi}+\eta_y\frac{\partial\eta_y}{\partial\eta}\right)\frac{\partial R}{\partial\eta}
\end{aligned}$$

Using previously defined expressions given by (11-210d), (11-211d), and (11-212e) repeated here for convenience, and additional definitions, the Laplacian becomes

$$\nabla^2 R = a_4\frac{\partial^2 R}{\partial\xi^2}+2c_5\frac{\partial^2 R}{\partial\xi\partial\eta}+b_4\frac{\partial^2 R}{\partial\eta^2}+g_1\frac{\partial R}{\partial\xi}+g_2\frac{\partial R}{\partial\eta}$$

where

$$\begin{aligned}
a_4 &= \xi_x^2+\xi_y^2\\
b_4 &= \eta_x^2+\eta_y^2\\
c_5 &= \xi_x\eta_x+\xi_y\eta_y\\
g_1 &= \xi_x\frac{\partial\xi_x}{\partial\xi}+\eta_x\frac{\partial\xi_x}{\partial\eta}+\xi_y\frac{\partial\xi_y}{\partial\xi}+\eta_y\frac{\partial\xi_y}{\partial\eta}\\
g_2 &= \xi_x\frac{\partial\eta_x}{\partial\xi}+\eta_x\frac{\partial\eta_x}{\partial\eta}+\xi_y\frac{\partial\eta_y}{\partial\xi}+\eta_y\frac{\partial\eta_y}{\partial\eta}
\end{aligned}$$

The second term $\nabla \nu_t \cdot \nabla R$ is first expanded as

$$\nabla \nu_t \cdot \nabla R = \frac{\partial \nu_t}{\partial x}\frac{\partial R}{\partial x} + \frac{\partial \nu_t}{\partial y}\frac{\partial R}{\partial y}$$

Now, using the Cartesian derivatives, one has

$$\left(\xi_x \frac{\partial \nu_t}{\partial \xi} + \eta_x \frac{\partial \nu_t}{\partial \eta}\right)\left(\xi_x \frac{\partial R}{\partial \xi} + \eta_x \frac{\partial R}{\partial \eta}\right) + \left(\xi_y \frac{\partial \nu_t}{\partial \xi} + \eta_y \frac{\partial \nu_t}{\partial \eta}\right)\left(\xi_y \frac{\partial R}{\partial \xi} + \eta_y \frac{\partial R}{\partial \eta}\right) =$$

$$\xi_x^2 \frac{\partial \nu_t}{\partial \xi}\frac{\partial R}{\partial \xi} + \xi_x \eta_x \frac{\partial \nu_t}{\partial \xi}\frac{\partial R}{\partial \eta} + \eta_x \xi_x \frac{\partial \nu_t}{\partial \eta}\frac{\partial R}{\partial \xi} + \eta_x^2 \frac{\partial \nu_t}{\partial \eta}\frac{\partial R}{\partial \eta}$$

$$+ \xi_y^2 \frac{\partial \nu_t}{\partial \xi}\frac{\partial R}{\partial \xi} + \xi_y \eta_y \frac{\partial \nu_t}{\partial \xi}\frac{\partial R}{\partial \eta} + \eta_y \xi_y \frac{\partial \nu_t}{\partial \eta}\frac{\partial R}{\partial \xi} + \eta_y^2 \frac{\partial \nu_t}{\partial \eta}\frac{\partial R}{\partial \eta}$$

$$= \left(\xi_x^2 + \xi_y^2\right)\frac{\partial \nu_t}{\partial \xi}\frac{\partial R}{\partial \xi} + (\xi_x \eta_x + \xi_y \eta_y)\frac{\partial \nu_t}{\partial \eta}\frac{\partial R}{\partial \eta}$$

$$+ (\xi_x \eta_x + \xi_y \eta_y)\frac{\partial \nu_t}{\partial \eta}\frac{\partial R}{\partial \xi} + (\eta_x^2 + \eta_y^2)\frac{\partial \nu_t}{\partial \eta}\frac{\partial R}{\partial \eta}$$

$$= a_4 \frac{\partial \nu_t}{\partial \xi}\frac{\partial R}{\partial \xi} + c_5\left(\frac{\partial \nu_t}{\partial \xi}\frac{\partial R}{\partial \eta} + \frac{\partial \nu_t}{\partial \eta}\frac{\partial R}{\partial \xi}\right) + b_4 \frac{\partial \nu_t}{\partial \eta}\frac{\partial R}{\partial \eta}$$

Now, consider the Baldwin-Barth turbulence model given by (21-87). The transformation of the convective term is exactly the same as the previous case, that is

$$\vec{V} \cdot \nabla R = u\frac{\partial R}{\partial x} + v\frac{\partial R}{\partial y} = U\frac{\partial R}{\partial \xi} + V\frac{\partial R}{\partial \eta}$$

The Laplacian in the diffusion term is first expanded in Cartesian coordinates and expressed as

$$\nabla^2 R = \frac{\partial^2 R}{\partial x^2} + \frac{\partial^2 R}{\partial y^2} = \frac{\partial}{\partial x}\left(\frac{\partial R}{\partial x}\right) + \frac{\partial}{\partial y}\left(\frac{\partial R}{\partial y}\right) = \frac{\partial A}{\partial x} + \frac{\partial B}{\partial y}$$

where

$$A = \frac{\partial R}{\partial x} \qquad \text{and} \qquad B = \frac{\partial R}{\partial y}$$

Now relations (21-88) and (21-89) are used to provide

$$\begin{aligned}
\nabla^2 R &= \xi_x \frac{\partial A}{\partial \xi} + \eta_x \frac{\partial A}{\partial \eta} + \xi_y \frac{\partial B}{\partial \xi} + \eta_y \frac{\partial B}{\partial \eta} \\
&= \xi_x \frac{\partial A}{\partial \xi} + \xi_y \frac{\partial B}{\partial \xi} + \eta_x \frac{\partial A}{\partial \eta} + \eta_y \frac{\partial B}{\partial \eta}
\end{aligned}$$

The second expression in the diffusion term $\nabla \cdot \nu_t \nabla R$ is written as

$$\nabla \cdot \nu_t \nabla R = \nabla \cdot \left(\nu_t \frac{\partial R}{\partial x} \vec{\imath} + \nu_t \frac{\partial R}{\partial y} \vec{\jmath} \right)$$

$$= \frac{\partial}{\partial x}\left(\nu_t \frac{\partial R}{\partial x}\right) + \frac{\partial}{\partial y}\left(\nu_t \frac{\partial R}{\partial y}\right) = \frac{\partial C}{\partial x} + \frac{\partial D}{\partial y}$$

where

$$C = \nu_t \frac{\partial R}{\partial x} \qquad \text{and} \qquad D = \nu_t \frac{\partial R}{\partial y}$$

In the generalized coordinates, one has

$$\nabla \cdot \nu_t \nabla R = \xi_x \frac{\partial C}{\partial \xi} + \eta_x \frac{\partial C}{\partial \eta} + \xi_y \frac{\partial D}{\partial \xi} + \eta_y \frac{\partial D}{\partial \eta}$$

$$= \xi_x \frac{\partial}{\partial \xi}\left[\nu_t \left(\xi_x \frac{\partial R}{\partial \xi} + \eta_x \frac{\partial R}{\partial \eta}\right)\right] + \eta_x \frac{\partial}{\partial \eta}\left[\nu_t \left(\xi_x \frac{\partial R}{\partial \xi} + \eta_x \frac{\partial R}{\partial \eta}\right)\right]$$

$$+ \xi_y \frac{\partial}{\partial \xi}\left[\nu_t \left(\xi_y \frac{\partial R}{\partial \xi} + \eta_y \frac{\partial R}{\partial \eta}\right)\right] + \eta_y \frac{\partial}{\partial \eta}\left[\nu_t \left(\xi_y \frac{\partial R}{\partial \xi} + \eta_y \frac{\partial R}{\partial \eta}\right)\right]$$

$$= \xi_x \frac{\partial \alpha}{\partial \xi} + \xi_y \frac{\partial \beta}{\partial \xi} + \eta_x \frac{\partial \alpha}{\partial \eta} + \eta_y \frac{\partial \beta}{\partial \eta}$$

where

$$\alpha = \nu_t \left(\xi_x \frac{\partial R}{\partial \xi} + \eta_x \frac{\partial R}{\partial \eta}\right)$$

$$\beta = \nu_t \left(\xi_y \frac{\partial R}{\partial \xi} + \eta_y \frac{\partial R}{\partial \eta}\right)$$

Finally, the following definitions are introduced which will be used in the formulation of FDE.

$$M^1 = -U \frac{\partial R}{\partial \xi} - V \frac{\partial R}{\partial \eta}$$

$$M^2 = -\frac{1}{Re_\infty} \frac{1}{\sigma_\epsilon} \left(\xi_x \frac{\partial \alpha}{\partial \xi} + \xi_y \frac{\partial \beta}{\partial \xi} + \eta_x \frac{\partial \alpha}{\partial \eta} + \eta_y \frac{\partial \beta}{\partial \eta}\right)$$

$$M^3 = \frac{1}{Re_\infty} 2 \left(\nu + \frac{\nu_t}{\sigma_\epsilon}\right) \left(\xi_x \frac{\partial A}{\partial \xi} + \xi_y \frac{\partial B}{\partial \xi} + \eta_x \frac{\partial A}{\partial \eta} + \eta_y \frac{\partial B}{\partial \eta}\right)$$

The last term of the Baldwin-Barth turbulence model includes a production term given by

$$(c_{\epsilon^2} f_2 - c_{\epsilon^1}) \sqrt{c_\mu D_1 D_2} \; \nu \widetilde{Re}_T \, S$$

where S in the computational space is given by (21-98).

APPENDIX K: The Transport Equation for the Turbulence Kinetic Energy

It is well established that turbulence in a flowfield is generated due to a variety of factors such as, for example, surface roughness, and it is convected and dissipated throughout the domain. To represent the above-mentioned physical aspect of turbulence in the development of a turbulence model, a transport equation is developed in terms of the turbulence kinetic energy. This transport equation which is derived from the Navier-Stokes equation is referred to as the *turbulence kinetic energy equation (TKE)*. The derivation of turbulence kinetic energy and the interpretation of terms in the equation are the focus of this section. The derivation of the equation will be performed in tensor notation. Recall that a repeated index in tensor notation indicates summation over that index.

The contintuity equation and the i-component of the momentum equation expressed in tensor notation are

$$\frac{\partial \rho}{\partial t} + \frac{\partial}{\partial x_j}(\rho u_j) = 0 \tag{K-1}$$

and

$$\rho \left(\frac{\partial u_i}{\partial t} + u_j \frac{\partial u_i}{\partial x_j} \right) = -\frac{\partial p}{\partial x_j} + \frac{\partial \tau_{ij}}{\partial x_j} \tag{K-2}$$

where

$$\tau_{ij} = \mu \left(\frac{\partial u_i}{\partial x_j} + \frac{\partial u_j}{\partial x_i} \right) - \delta_{ij} \frac{2}{3} \mu \frac{\partial u_k}{\partial x_k} \tag{K-3}$$

The Reynolds averaged continuity and the i-component of momentum equations are given by

$$\frac{\partial \bar{\rho}}{\partial t} + \frac{\partial}{\partial x_j}(\bar{\rho}\,\bar{u}_j) + \frac{\partial}{\partial x_j}(\overline{\rho' u_j'}) = 0 \tag{K-4}$$

and

$$\bar{\rho} \frac{\partial \bar{u}_i}{\partial t} + \overline{\rho' \frac{\partial u_i'}{\partial t}} + \left(\bar{\rho}\,\bar{u}_j + \overline{\rho' u_j'} \right) \frac{\partial \bar{u}_i}{\partial x_j}$$

$$+\frac{\partial}{\partial x_j}\left(\bar{\rho}\,\overline{u_i' u_j'}\right)+\frac{\partial}{\partial x_j}\left(\bar{u}_j\,\overline{\rho' u_j'}\right)+\frac{\partial}{\partial x_j}\left(\overline{\rho' u_i' u_j'}\right)=$$

$$=-\frac{\partial \bar{p}}{\partial x_i}+\frac{\partial \bar{\tau}_{ij}}{\partial x_j} \tag{K-5}$$

To proceed with the derivation of a TKE equation, multiply the continuity equation by $1/2\ u_i^2$ to obtain

$$\frac{1}{2}u_i^2\left[\frac{\partial \rho}{\partial t}+\frac{\partial}{\partial x_j}(\rho\, u_j)\right]=0 \tag{K-6}$$

Multiply the i-component of the momentum equation by u_i to obtain

$$\rho\, u_i\left(\frac{\partial u_i}{\partial t}+u_j\frac{\partial u_i}{\partial x_j}\right)=-u_i\frac{\partial p}{\partial x_j}+u_i\frac{\partial \tau_{ij}}{\partial x_j} \tag{K-7}$$

Note that the second term in Equation (K-7) can be rearranged as

$$\rho\, u_i\, u_j\frac{\partial u_i}{\partial x_j}=\rho\, u_j\left(\frac{1}{2}\frac{\partial u_i^2}{\partial x_j}\right)$$

Thus, Equation (K-7) is written as

$$\rho u_i\frac{\partial u_i}{\partial t}+\frac{1}{2}\rho u_j\frac{\partial}{\partial x_j}(u_i^2)=-u_i\frac{\partial p}{\partial x_i}+u_i\frac{\partial \tau_{ij}}{\partial x_j} \tag{K-8}$$

Add Equation (K-8) to Equation (K-1) to obtain

$$\frac{\partial}{\partial t}\left(\rho\frac{u_i^2}{2}\right)+\frac{\partial}{\partial x_j}\left(\rho\, u_j\, u_i^2\right)=-u_i\frac{\partial p}{\partial x_i}+u_i\frac{\partial \tau_{ij}}{\partial x_j} \tag{K-9}$$

Now substitute the sum of average and fluctuating quantities for the instantaneous quantities, so that Equation (K-9) becomes

$$\frac{\partial}{\partial t}\left[\frac{1}{2}(\bar{\rho}+\rho')(\bar{u}_i+u_i')^2\right]+\frac{\partial}{\partial x_j}\left[\frac{1}{2}(\bar{\rho}+\rho')(\bar{u}_j+u_j')(\bar{u}_i+u_i')^2\right]$$

$$-(\bar{u}_i+u_i')\frac{\partial}{\partial x_i}(\bar{p}+p')+(\bar{u}_i+u_i')\frac{\partial}{\partial x_j}(\bar{\tau}_{ij}+\tau_{ij}')$$

Taking the time average following the rules set by (17-2), one obtains

$$\frac{1}{2}\frac{\partial}{\partial t}\left(\bar{\rho}\,\bar{u}_i^2+\bar{\rho}\,\overline{u_i'^2}+\overline{\rho' u_i'^2}+2\bar{u}_i\,\overline{\rho' u_i'}\right)$$

$$+\frac{1}{2}\frac{\partial}{\partial x_j}\left[\bar{\rho}\left(\bar{u}_j\bar{u}_i^2+\bar{u}_j\overline{u_i'^2}\right)+\bar{u}_j\overline{\rho' u_i'^2}+2\bar{u}_i\bar{u}_j\overline{\rho' u_i'}\right.$$

$$+ \bar{\rho}\,\overline{u_j' u_i'^2} + 2\,\bar{\rho}\,\bar{u}_i\,\overline{u_i' u_j'} + \bar{u}_i^2\,\overline{\rho' u_j'} + \overline{\rho' u_j' u_i'^2} + 2\,\bar{u}_i\,\overline{\rho' u_i' u_j'}\Big]$$

$$= \quad -\bar{u}_i \frac{\partial \bar{p}}{\partial x_i} - \overline{u_i' \frac{\partial p'}{\partial x_i}} + \bar{u}_i \frac{\partial \bar{\tau}_{ij}}{\partial x_j} + \overline{u_i' \frac{\partial \tau_{ij}'}{\partial x_j}} \qquad \text{(K-10)}$$

Now multiply the momentum equation given by Equation (K-5) by $\bar{u}_i$.

$$\bar{u}_i \left(\bar{\rho}\,\frac{\partial \bar{u}_i}{\partial t} + \overline{\rho'\,\frac{\partial u_i'}{\partial t}} \right) + \bar{u}_i \left(\bar{\rho}\,\bar{u}_j + \overline{\rho' u_j'} \right) \frac{\partial \bar{u}_i}{\partial x_j}$$

$$+ \bar{u}_i \frac{\partial}{\partial x_j} \left(\bar{\rho}\,\overline{u_i' u_j'} \right) + \bar{u}_i \frac{\partial}{\partial x_j} \left(\bar{u}_j \overline{\rho' u_i'} \right) + \bar{u}_i \frac{\partial}{\partial x_j} \left(\overline{\rho' u_i' u_j'} \right)$$

$$= \quad -\bar{u}_i \frac{\partial \bar{p}}{\partial x_i} + \bar{u}_i \frac{\partial \bar{\tau}_{ij}}{\partial x_j} \qquad \text{(K-11)}$$

Subtract Equation (K-11) from Equation (K-10) and collect terms to obtain

$$\frac{1}{2}\,\bar{u}_i^2 \frac{\partial \bar{\rho}}{\partial t} + \frac{1}{2} \frac{\partial}{\partial t} \left(\bar{\rho}\,\overline{u_i'^2} \right) + \frac{1}{2} \frac{\partial}{\partial t} \left(\overline{\rho' u_i'^2} \right) + \overline{\rho' u_i'}\,\frac{\partial \bar{u}_i}{\partial t}$$

$$+ \frac{1}{2} \frac{\partial}{\partial x_j} \left(\bar{\rho}\,\bar{u}_j \bar{u}_i^2 \right) - \bar{p}\,\bar{u}_i \bar{u}_j \frac{\partial \bar{u}_i}{\partial x_j}$$

$$+ \frac{1}{2} \frac{\partial}{\partial x_j} \left(\bar{u}_i^2 \overline{\rho' u_j'} \right) - \bar{u}_i \overline{\rho' u'}_j \frac{\partial \bar{u}_i}{\partial x_j}$$

$$+ \frac{\partial}{\partial x_j} \left(\bar{\rho}\,\bar{u}_i \overline{u_i' u_j'} \right) - \bar{u}_i \frac{\partial}{\partial x_j} \left(\bar{\rho}\,\overline{u_i' u_j'} \right)$$

$$+ \frac{\partial}{\partial x_j} \left(\bar{u}_i\,\bar{u}_j \overline{\rho' u_i'} \right) - \bar{u}_i \frac{\partial}{\partial x_j} \left(\bar{u}_j\,\overline{\rho' u_i'} \right)$$

$$+ \frac{\partial}{\partial x_j} \left(\bar{u}_i \overline{\rho' u_i' u_j'} \right) - \bar{u}_i \frac{\partial}{\partial x_j} \left(\overline{\rho' u_i' u_j'} \right)$$

$$+ \frac{1}{2} \frac{\partial}{\partial x_j} \left[\bar{\rho}\,\bar{u}_j \overline{u_i'^2} + \bar{u}_j \overline{\rho' u_i'^2} + \bar{\rho}\,\overline{u_j' u_i'^2} + \overline{\rho' u_j' u_i'^2} \right]$$

$$= \quad -\overline{\bar{u}_i' \frac{\partial p'}{\partial x_i}} + \overline{u_i' \frac{\partial \tau_{ij}'}{\partial x_j}} \qquad \text{(K-12)}$$

The fifth and sixth terms can be combined as follows

$$\frac{1}{2}\frac{\partial}{\partial x_j}\left(\bar{\rho}\,\bar{u}_j\bar{u}_i^2\right) - \bar{\rho}\,\bar{u}_i\bar{u}_j\frac{\partial \bar{u}_i}{\partial x_j} = \frac{1}{2}\bar{u}_i^2\frac{\partial}{\partial x_j}\left(\bar{\rho}\,\bar{u}_j\right)$$

The remaining terms can be combined in a similar fashion, and Equation (K-12) is written as

$$\frac{1}{2}\bar{u}_i^2\frac{\partial \bar{\rho}}{\partial t} + \frac{1}{2}\frac{\partial}{\partial t}\left(\bar{\rho}\,\overline{u_i'^2}\right) + \frac{1}{2}\frac{\partial}{\partial t}\left(\overline{\rho' u_i'^2}\right) + \overline{\rho' u_i'}\frac{\partial \bar{u}_i}{\partial t}$$

$$+\frac{1}{2}\bar{u}_i^2\frac{\partial}{\partial x_j}\left(\bar{\rho}\,\bar{u}_j\right) + \bar{\rho}\,\overline{u_i'u_j'}\frac{\partial \bar{u}_i}{\partial x_j} + \frac{1}{2}\bar{u}_i^2\frac{\partial}{\partial x_j}\left(\overline{\rho' u_j'}\right)$$

$$+\overline{\rho' u_i'}\bar{u}_j\frac{\partial \bar{u}_i}{\partial x_j} + \overline{\rho' u_i' u_j'}\frac{\partial \bar{u}_i}{\partial x_j}$$

$$+\frac{1}{2}\frac{\partial}{\partial x_j}\left[\bar{\rho}\,\bar{u}_j\overline{u_i'^2} + \bar{u}_j\,\overline{\rho' u_i'^2} + \bar{\rho}\,\overline{u_j' u_i'^2} + \overline{\rho' u_j' u_i'^2}\right]$$

$$= \quad -\overline{\bar{u}_i'\frac{\partial p'}{\partial x_i}} + \overline{u_i'\frac{\partial \bar{\tau}_{ij}'}{\partial x_j}} \qquad \text{(K-13)}$$

Observe that the sum of the first and fifth terms is zero, due to continuity, and one can write

$$-\overline{u_i'\frac{\partial p'}{\partial x_i}} = -\frac{\partial}{\partial x_i}\left(\overline{p' u_i'}\right) + \overline{p'\frac{\partial u_i'}{\partial x_i}} \qquad \text{(K-14)}$$

Now define the turbulent kinetic energy k as

$$k = \frac{1}{2}\left(u_i'^2\right) \qquad \text{(K-15)}$$

Finally, Equation (K-13) is written as

$$\frac{\partial}{\partial t}\left(\bar{\rho}\,k\right) + \frac{\partial}{\partial x_j}\left(\bar{\rho}\,\bar{u}_j\,k + \overline{\rho' u_j' k}\right) = -\frac{\partial}{\partial x_i}\left(\overline{p' u_i'}\right)$$

$$-\frac{\partial}{\partial x_j}\left(\bar{\rho}\,\overline{u_j' k}\right) - \frac{\partial}{\partial x_j}\left(\bar{u}_j\,\overline{\rho' k}\right) - \bar{\rho}\,\overline{u_i' u_j'}\frac{\partial \bar{u}_i}{\partial x_j}$$

$$-\overline{\rho' u_i' u_j'}\frac{\partial \bar{u}_i}{\partial x_j} - \overline{\rho' u_i'}\,\bar{u}_j\frac{\partial \bar{u}_i}{\partial x_j} + \overline{p'\frac{\partial u_i'}{\partial x_i}} + \overline{u_i'\frac{\partial \tau_{ij}'}{\partial x_j}} \qquad \text{(K-16)}$$

This is the turbulence kinetic energy equation.

Recall that the shear stress is given by

$$\tau_{ij} = \mu\left(\frac{\partial u_i}{\partial x_j} + \frac{\partial u_j}{\partial x_i}\right) - \delta_{ij}\,\frac{2}{3}\,\mu\,\frac{\partial u_k}{\partial x_k} \qquad \text{(K-17)}$$

Furthermore, one can write

$$\overline{u_i'\frac{\partial \tau_{ij}'}{\partial x_j}} = \frac{\partial}{\partial x_j}\left(\overline{u_i'\tau_{ij}'}\right) - \overline{\tau_{ij}'\frac{\partial u_i'}{\partial x_j}} \qquad \text{(K-18)}$$

Now, an alternate form of the TKE equation can be written by substitution of (K-17) and (K-18) into (K-16), as follows

$$\frac{\partial}{\partial t}\left(\bar{\rho}\, k\right) + \frac{\partial}{\partial x_j}\left(\bar{\rho}\,\bar{u}_j\, k + \overline{\rho' u_j' k}\right) = -\frac{\partial}{\partial x_i}\left(\overline{p' u_i'}\right)$$

$$-\frac{\partial}{\partial x_j}\left(\bar{\rho}\,\overline{u_j' k}\right) - \frac{\partial}{\partial x_j}\left(\bar{u}_j \overline{\rho' k}\right) - \bar{\rho}\,\overline{u_i' u_j'}\,\frac{\partial \bar{u}_i}{\partial x_j}$$

$$-\left(\bar{u}_j \overline{\rho' u_i'} + \overline{\rho' u_i' u_j'}\right)\frac{\partial \bar{u}_i}{\partial x_j}$$

$$+\frac{\partial}{\partial x_j}\left\{\overline{u_i'\left[\mu\left(\frac{\partial u_i'}{\partial x_j} + \frac{\partial u_j'}{\partial x_i}\right) - \delta_{ij}\frac{2}{3}\,\mu\,\frac{\partial u_k'}{\partial x_k}\right]}\right\}$$

$$= -\overline{\left\{\mu\left(\frac{\partial u_i'}{\partial x_j} + \frac{\partial u_j'}{\partial x_i}\right) - \delta_{ij}\frac{2}{3}\mu\,\frac{\partial u_k'}{\partial x_k}\right\}\frac{\partial u_i'}{\partial x_j}} + \overline{p'\frac{\partial u_i'}{\partial x_i}} \qquad \text{(K-19)}$$

The interpretation of the terms in the TKE equation given by (K-19) is as follows.

I. The rate of change of TKE which is composed of a local change and a convective term is

$$\frac{\partial}{\partial t}\left(\bar{\rho}\, k\right) + \frac{\partial}{\partial x_j}\left(\bar{\rho}\,\bar{u}_j\, k + \overline{\rho' u_j'\, k}\right) \qquad \text{(K-20)}$$

II. The diffusion term includes the following

$$-\frac{\partial}{\partial x_i}\left(\overline{p' u_i'}\right) - \frac{\partial}{\partial x_j}\left(\bar{\rho}\,\overline{u_j'\, k}\right) - \frac{\partial}{\partial x_j}\left(\bar{u}_j\,\overline{\rho' k}\right) \qquad \text{(K-21)}$$

and represents diffusion of turbulence due to molecular transport processes.

III. The production term is

$$- \bar{\rho}\, \overline{u_i' u_j'}\, \frac{\partial \bar{u}_i}{\partial x_j} - \left(\bar{u}_j\, \overline{\rho' u_i'} + \overline{\rho' u_i' u_j'} \right) \frac{\partial \bar{u}_i}{\partial x_j} \tag{K-22}$$

and it represents the rate of transfer of kinetic energy of the mean flow to turbulence.

IV. The work done by the turbulent viscous stresses is given by

$$\overline{\frac{\partial}{\partial x_j} \left\{ u_i' \left[\mu \left(\frac{\partial u_i'}{\partial x_j} + \frac{\partial u_j'}{\partial x_i} \right) - \delta_{ij}\, \frac{2}{3} \mu \frac{\partial u_k'}{\partial x_k} \right] \right\}} \tag{K-23}$$

V. The turbulence viscous dissipation is

$$- \overline{\left\{ \mu \left(\frac{\partial u_i'}{\partial x_j} + \frac{\partial u_j'}{\partial x_i} \right) - \delta_{ij}\, \frac{2}{3} \mu \frac{\partial u_k'}{\partial x_k} \right\} \frac{\partial u_i'}{\partial x_j}} \tag{K-24}$$

and finally

VI. The pressure dilatation is

$$\overline{p' \frac{\partial u_i'}{\partial x_i}} \tag{K-25}$$

Typically the fluctuating correlations involving density fluctuations are dropped. Furthermore, the Boussinesq assumption is invoked, and the terms defined by (K-20) through (K-25) are written as

I. Rate of change of TKE is

$$\frac{\partial}{\partial t} (\bar{\rho}\, k) + \frac{\partial}{\partial x_j} (\bar{\rho}\, \bar{u}_j\, k) \tag{K-26}$$

Note that this expression can be modified as follows

$$\frac{\partial}{\partial t} (\bar{\rho}\, k) + \frac{\partial}{\partial x_j} (\bar{\rho}\, \bar{u}_j\, k) = \bar{\rho}\, \frac{\partial k}{\partial t} + k\, \frac{\partial \bar{\rho}}{\partial t} + \rho\, \bar{u}_j\, \frac{\partial k}{\partial x_j} + k\, \frac{\partial}{\partial x_j} (\bar{\rho}\, \bar{u}_j)$$

$$= \bar{\rho} \left(\frac{\partial k}{\partial t} + u_j\, \frac{\partial k}{\partial x_j} \right) + k \left[\frac{\partial \rho}{\partial t} + \frac{\partial}{\partial x_j} (\bar{\rho}\, \bar{u}_j) \right]$$

$$= \bar{\rho} \left(\frac{\partial k}{\partial t} + u_j\, \frac{\partial k}{\partial x_j} \right) = \bar{\rho}\, \frac{dk}{dt} \tag{K-27}$$

where continuity is used to set the bracket term equal to zero.

II. The diffusion term is written as

$$\frac{\partial}{\partial x_j}\left(\frac{\mu_t}{\sigma_k}\frac{\partial k}{\partial x_j}\right) \tag{K-28}$$

III. The production is denoted by P_k and is given by

$$P_k = -\bar{\rho}\,\overline{u_i' u_j'}\,\frac{\partial \bar{u}_i}{\partial x_j} = \tau_{ij}\,\frac{\partial \bar{u}_i}{\partial x_j} \tag{K-29}$$

where τ_{ij} now represents the turbulent shear stress and is given by

$$\tau_{ij} = \mu_t\left(\frac{\partial \bar{u}_i}{\partial x_j} + \frac{\partial \bar{u}_j}{\partial x_i} - \frac{2}{3}\frac{\partial u_k}{\partial x_k}\,\delta_{ij}\right) - \delta_{ij}\,\frac{2}{3}\,\bar{\rho}\,k \tag{K-30}$$

or

$$\tau_{ij} = \left\{\mu_t\left(\frac{\partial \bar{u}_i}{\partial x_j} + \frac{\partial \bar{u}_j}{\partial x_i} - \frac{2}{3}\frac{\partial u_k}{\partial x_k}\,\delta_{ij}\right) - \delta_{ij}\,\frac{2}{3}\,\bar{\rho}\,k\right\}\frac{\partial \bar{u}_i}{\partial x_j} \tag{K-31}$$

IV. The work done by the turbulent viscous stress is

$$\frac{\partial}{\partial x_j}\left(\mu\,\frac{\partial k}{\partial x_j}\right) \tag{K-32}$$

V. The turbulent viscous dissipation, which is typically referred to as *dissipation*, is

$$-\,\mu\,\overline{\frac{\partial u_i'}{\partial x_j}\frac{\partial u_i'}{\partial x_j}} = -\bar{\rho}\,\epsilon \tag{K-33}$$

VI. The pressure dilatation term is usually dropped. Finally, the TKE equation is expressed as follows

$$\frac{\partial}{\partial t}\,(\bar{\rho}\,k) + \frac{\partial}{\partial x_j}\,(\bar{\rho}\,\bar{u}_j\,k) = \frac{\partial}{\partial x_j}\left[\left(\mu + \frac{\mu_t}{\sigma_k}\right)\frac{\partial k}{\partial x_j}\right] + P_k - \bar{\rho}\epsilon \tag{K-34}$$

REFERENCES

[20-1] Robinson, S.K. "Coherent Motions in the Turbulent Boundary Layer," *Annual Review Fluid Mechanics,* 23, pp. 601–639, 1991.

[20-2] White, F.M., "Viscous Fluid Flow," McGraw-Hill Book Company, 1974.

[20-3] Panton, R.L., "Incompressible Flow," John Wiley and Sons, 1984.

[20-4] Landahl, M.T., and Mollo-Christensen, E., "Turbulence and Random Processes in Fluid Mechanics," Second Edition, Cambridge University Press, 1992.

[20-5] Squire, H.B., "On the Stability of the Three-Dimensional Disturbances of Viscous Flow Between Parallel Walls," Proc. R. Soc. London, Ser. A, Vol. 142, 1933, pp. 621–628.

[20-6] Blackwelder, R.F., "Analogies Between Transitional and Turbulent Boundary Layers," *Physics of Fluids,* Vol. 26, No. 10, 1983, pp. 2807–2815.

[20-7] Myose, R.Y., private communication, 1998.

[20-8] Kim, H.T., Kline, S.J., and Reynolds, W.C., "The Production of Turbulence Near a Smooth Wall in a Turbulent Boundary Layer," *J. Fluid Mechanics,* Vol. 50, 1971, pp. 133–160.

[20-9] Blackwelder, R.F., and Kovasznzy, L.S., "Time Scales and Correlations in a Turbulent Boundary Layer," *Physics of Fluids,*, Vol15, No. 9, 1972, pp. 1545–1554.

[20-10] Myose, R.Y., and Blackwelder, R.F., "On the Role of the Outer Region in the Turbulent Boundary Layer Bursting Process," *J. Fluid Mechanics,* Vol. 259, 1994, pp. 345–373.

[21-1] Wilcox, D. C., "Turbulence Modeling for CFD," DCW Industries, Inc., 1993.

[21-2] ___"Turbulence in Compressible Flows," AGARD Report 819, June 1997.

[21-3] Chen, C-J., and Jaw, S-Y., "Fundamentals of Turbulence Modeling," Taylor and Francis, 1998.

[21-4] Libby, P. A., "Introduction to Turbulence," Taylor and Francis, 1996.

[21-5] Marvin, J. G., "Turbulence Modeling for Computational Aerodynamics," AIAA-82-0164, 1982.

[21-6] Kral, L. D., Mani, M., and Ladd, J. A., "On the Application of Turbulence Models for Aerodynamic and Propulsion Flowfields," AIAA-96-0564, January 1996.

[21-7] Rumsey, C. L., and Natsa, V. N., "Comparison of the Predictive Capabilities of Several Turbulence Models," *Journal of Aircraft,* Vol. 32, No. 3, May–June 1995.

[21-8] Kern, S., "Evaluation of Turbulence Models for High-Lift Military Airfoil Flowfields," AIAA-96-0057, January 1996.

[21-9] Baldwin, B. S. and Lomax, H., "Thin-Layer Approximation and Algebraic Model for Separated Turbulent Flows," AIAA-78-257, 1978.

[21-10] Granville, P. S., "Baldwin-Lomax Factors for Turbulent Boundary Layers in Pressure Gradients," *AIAA Journal,* Vol. 25, No. 12, December 1987.

[21-11] Baldwin, B. S., and Barth, T. J., "A One-Equation Turbulence Transport Model for High Reynolds Number Wall-Bounded Flows," NASA-TM-102847, August 1990.

[21-12] Spalart, P. R., and Allmaras, S. R., "A One-Equation Turbulence Model for Aerodynamics Flows," AIAA-92-0439, January 1992.

[21-13] Jones, W. P., and Launder, B. E., "The Prediction of Laminarization with a Two-Equation Model of Turbulence," *International Journal of Heat and Mass Transfer,* Vol. 15, 1972, pp. 301–314.

[21-14] Jones, W. P., and Launder, B. E., "The Calculation of Low-Reynolds-Number Phenomena with a Two-Equation Model of Turbulence," *International Journal of Heat and Mass Transfer,* Vol. 16, 1973, pp. 1119–1130.

[21-15] Hoffman, G. H., "Improved Form of the Low Reynolds Number k-ϵ Turbulence Model," *Physics of Fluids,* Vol. 18, No. 3, March 1975, pp. 309–312.

[21-16] Nagano, Y., and Hishida, M., "Improved Form of the k-ϵ Model for Wall Turbulent Shear Flows," *Journal of Fluids Engineering,* Vol. 109, June 1987, pp. 156–160.

[21-17] Wilcox, D. C., "Reassessment of the Scale-Determining Equation for Advanced Turbulence Models," *AIAA Journal,* Vol. 26, November 1988, pp. 1299–1310.

[21-18] Menter, F. R., "Assessment of Higher Order Turbulence Models for Complex Two- and Three-Dimensional Flowfields," NASA-TM-103944, September 1992.

[21-19] Unalmis, O. H., and Dolling, D. S., "Experimental Study of Causes of Unsteadiness of Shock-Induced Turbulent Separation," *AIAA Journal,* Vol. 36, No. 3, March 1998, pp. 371–378.

[21-20] Dolling, D. S., and Murphy, M. T., "Unsteadiness of the Separation Shock Wave Structure in a Supersonic Compression Ramp Flowfield," *AIAA Journal,* Vol. 21, No. 12, December 1983, pp. 1628–1634.

[21-21] Brusniak, L., and Dolling, D. S., "Engineering Estimation of Fluctuating Loads in Shock Wave/Turbulent Boundary-Layer Interaction," *AIAA Journal,* Vol. 34, No. 12, December 1996, pp. 2554–2561.

[21-22] Settles, G. S., "An Experimental Study of Compressible Turbulent Boundary Layer Separation at High Reynolds Numbers," PhD Dissertation, Princeton University, 1975.

[21-23] Hoffmann, K. A., Suzen, Y. B., and Papadakis, M., "Numerical Computation of High Speed Exhaust Flows," AIAA-95-0758, January 1995.

[21-24] Seiner, J. M., Norum, T. D., "Experiments of Shock Associated Noise on Supersonic Jets," AIAA-79-1526, July 1979.

[21-25] Cebeci, T. and Bradshaw, P., "Physical and Computational Aspects of Convective Heat Transfer," Springer-Verlag, 1984.

[21-26] Kays, W. M. and Crawford, M. E., "Convective Heat and Mass Transfer," Second edition, McGraw-Hill, 1980.

[21-27] Landahl, M. T., and Mollo-Christensen, E., "Turbulence and Random Processes in Fluid Mechanics," Second Edition, Cambridge University Press, 1992.

[23-1] Leonard, A., "Energy Cascade in Large-Eddy Simulations of Turbulent Fluid Flows," *Advances in Geophysics*, Vol. 18A, pp. 237–248, 1974.

[23-2] Smagorinsky, J., "General Circulation Experiments with Primitive Equations. I. The Basic Experiment," *Monthly Weather Review*, Vol. 91, pp. 99–164, 1963.

[23-3] Germano, M., Piomelli, U., Moin, P., and Cabot, W. H., "A Dynamic Subgrid-Scale Eddy Viscosity Model," *Phys. Fluids A.*, Vol. 3, pp. 1760–1765, 199?.

[23-4] Lilly, D. K., *Phys. Fluids A.*, Vol. 4, p. 633, 1992.

[23-5] Lele, S. K., "Direct Numerical Simulation of Compressible Free Shear Flows," AIAA-89-0376, January 1989.

[23-6] Liu, Z., Zhao, W., and Liu, C., "Direct Numerical Simulation of Transition in a Subsonic Airfoil Boundary Layer," AIAA-97-0752, January 1997.

[23-7] Liu, Z., Zhao, W., Xiong, G., and Liu, C., "Direct Numerical Simulation of Flow Transition in High-Speed Boundary Layers Around Airfoils," AIAA-97-0753, January 1997.

[23-8] Rai, M. M., Gatski, T. B., and Erlebacher, G., "Direct Simulation of Spatially Evolving Compressible Turbulent Boundary Layers," AIAA-95-0583, January 1995.

[23-9] Sumitari, Y., and Kasagi, N., "Direct Numerical Simulation of Turbulent Transport with Uniform Wall Injection and Suction," *AIAA Journal*," Vol. 33, No. 7, pp. 1220–1228, July 1995.

[23-10] Hatay, F. F., and Biringen, S., "Direct Numerical Simulation of Low-Reynolds Number Supersonic Turbulent Boundary Layers," AIAA-95-0581, January 1995.

[23-11] Adams, N. A., "Direct Numerical Simulation of Turbulent Supersonic Boundary Layer Flow," *Advances in DNS/LES*, Liu, C., Liu, Z., and Sakell, L., editors, Greyden Press, pp. 29–40, 1997.

INDEX

The numbers 1, 2 or 3 preceding the page numbers refer to Volumes 1, 2 and 3, respectively.